Battery Technology Crash Course

Slobodan Petrovic

Battery Technology Crash Course

A Concise Introduction

 Springer

Slobodan Petrovic
Electrical Engineering and Renewable Energy
Oregon Institute of Technology
Happy Valley, OR, USA

ISBN 978-3-030-57271-6   ISBN 978-3-030-57269-3   (eBook)
https://doi.org/10.1007/978-3-030-57269-3

Cover Illustration Credit: Kevin Hudson.

This Springer imprint is published by the registered company Springer Nature Switzerland AG
The registered company address is: Gewerbestrasse 11, 6330 Cham, Switzerland

Preface

Batteries look simple on the outside. Most people look at a packaged battery and see a closed system that delivers energy on demand. The system does not communicate with its environment except electrically. It delivers electrical energy on demand, and for rechargeable batteries, it accepts electrical energy during charging. Not everyone asks themselves what is inside a battery or how a battery works. Yet there is a fascinating system inside a battery—workings of a complex organization with multifunctional components and reactions with the occasional mysterious outcomes. For researchers, those who develop and test batteries, they represent a challenge: a constant quest for improvement, for better material, for a favorable sequence of events during manufacturing, or for a new structure that might prove to advance one small part of the overall performance. By stretching the limits of chemistry, physics, material science, and most importantly, the overlapping boundaries between these fields, batteries have been changing for over 200 years, beginning with the invention of the voltaic pile, which was simply a series of copper and zinc disks separated by brine.

For the average person, even an engineer, a battery is a useful source of electricity that can be used to power many types of devices. In many situations, batteries are designed into a system with only minimal regard to their properties or the evaluation of their compatibility with the device the batteries are powering. Those who decide to venture a little deeper may study the main battery properties and consider more complex relationships. Still, the inner workings of a battery usually remain enigmatic for many. It has often been suspected that the basic challenge for many students arises from the fact that batteries are electrochemical devices and that the processes in batteries must be ultimately understood in terms of the interface between the electrical phenomena and chemical reactions. This has proven to be difficult not only for students of engineering but also for anyone in the technical or business community who wishes to gain an understanding of battery principles, performance, and use. While there are many extraordinary texts on batteries available, most of them are too long and too technically detailed for a reader generally inexperienced in the field. These fine books usually require a long course of study before a satisfactory level of understanding is gained. As a result, it appeared that a fundamental text was needed, which enables a rapid course of learning toward understanding the scientific and engineering principles involved in modern batteries.

I have discovered over many years of teaching that engineering students master the field of batteries better when taking only one course, if they are presented with the fundamental concepts first and then given an opportunity to gain an understanding of the electrochemical processes inside the batteries in conceptual principles instead of specific results. Adding additional layers of knowledge to basic understanding involves applying multi-level reasoning about the effects of interior and exterior parameters on battery reactions. The goal in the education of engineering students is to develop an understanding of trends in battery behavior under the influence of overlapping factors. This is the main motivation behind this textbook—a gradual and layered introduction into the difficult science and technology of batteries, followed by a chronological look at different battery types with the intention of establishing the reader's ability to confidently select an appropriate battery for an application and conditions of use. Ultimately, the desire to facilitate the learning of fundamental principles of batteries led to this crash course suitable for all engineers, analysts, economists, investment bankers, and other busy professionals who wish to rapidly gain a necessary understanding of batteries.

Happy Valley, OR   Slobodan Petrovic

Acknowledgment

My thanks to Laura Polk for her assistance with getting this book to the publisher on time.

Contents

Introduction

1

1.1 Galvanic Cells

Batteries are one type of galvanic cells. Galvanic cells are electrochemical devices that produce electricity in spontaneous reactions when their electrodes are connected through a load and in contact with an electrolyte. Besides batteries, fuel cells are the other main type of galvanic cell.

The difference between batteries and fuel cells is that batteries can deliver a limited and predetermined amount of electricity based on the finite quantity of reactants in their enclosed casing, while fuel cells operate as long as reactants (fuel and oxidant) are supplied from external sources. Some fuel cells operate with one electrode, the cathode, utilizing oxygen from air and have only the fuel for the anode supplied.

The distinction between batteries and fuel cells can be defined as the ability of certain batteries to be recharged by applying voltage to the cell and reversing the battery reaction (i.e., rechargeable batteries). Fuel cells are usually not considered "rechargeable" although so-called reversible fuel cells can operate in the reverse mode as electrolyzers. The schematic principles of batteries and fuel cells are shown in Fig. 1.1. Note reactant inlets and outlets on both sides of the fuel cell sketch on the right.

It is worth mentioning that flow batteries, an additional type of galvanic cell, combine principles of both batteries and fuel cells. They are batteries because of the recharging capability, but closer to fuel cells because the reactants are supplied from outside reservoirs. When all the active material, mixed with the electrolyte is reacted, the reaction switches from discharge to charge.

Metal–air batteries are devices that combine one traditional battery metal anode and a cathode that operates on oxygen from air, just like the majority of fuel cell cathodes (Fig. 1.2).

© The Editor(s) (if applicable) and The Author(s), under exclusive license to
Springer Nature Switzerland AG 2021
S. Petrovic, *Battery Technology Crash Course*,
https://doi.org/10.1007/978-3-030-57269-3_1

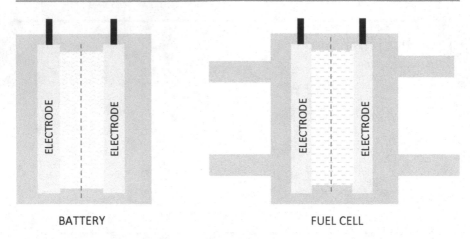

Fig. 1.1 Schematic representation of a battery as a closed system (left) and a fuel cell (right)

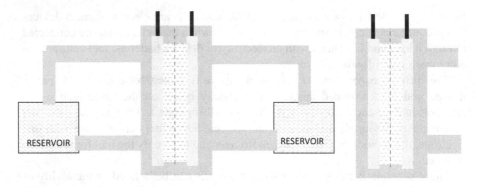

Fig. 1.2 Schematic representation of a flow battery (left) and a metal–air battery (right)

1.2   Types of Batteries

Batteries can be classified based on many different factors such as reversibility of reaction, type of electrolyte, principle of operation, or application. Two main battery types are primary batteries and secondary or rechargeable batteries.

The primary, non-rechargeable batteries utilize electrochemical reactions during discharge that cannot be effectively reversed by applying voltage to a cell. Consequently, they are discarded after being discharged once. Primary batteries make up roughly 20% of the overall battery market, but they have some distinct advantages compared with rechargeable batteries, such as higher energy content per weight and volume, longer shelf lives, and full capacity on demand. The common cylindrical cells, such as double A, triple A, 9-V, and button-type are mainly used for portable electronic devices, lighting, cameras, toys, and similar items. Select primary batteries with higher capacity are used for military applications.

Important primary batteries include zinc–carbon cell or so-called dry cell; magnesium and aluminum batteries; alkaline manganese dioxide, which is the most popular primary battery; mercuric oxide and silver oxide; and finally zinc–air and lithium batteries. The study of primary batteries is not the topic of this textbook.

In secondary, rechargeable batteries, applying a voltage can reverse the reactions that take place during the discharge and drive the battery back to a charged state from which it can deliver electrical energy. Depending on the battery type, charge–discharge cycle can be repeated many times, from about 500 cycles for popular lead–acid batteries to over 10,000 cycles for typical flow batteries. As a result of their ability to be recharged, secondary batteries can function as electrical energy storage devices, also called accumulators. The most widely used rechargeable batteries described in this text include lead–acid, nickel-based batteries, and lithium–ion batteries.

Another battery type has been created to minimize problems that some batteries have with short shelf life and high rates of self-discharge. In these so-called reserve batteries, one component, e.g., an electrode, is inserted into the battery at the time of demand, which makes the battery fully charged when needed. Alternatively, with so-called thermal batteries, the battery becomes operational only after heating the electrolyte to make it conductive.

Finally, flow batteries are a special version of a battery concept where active mass is not in the electrodes, but it is mixed with the electrolyte, stored in the reservoirs, and recirculated through the cell. The main distinguishing feature of flow batteries is that sites where active mass is stored are different from sites where electrochemical reactions occur. This site separation prevents degradation of active material and also enables its easy replacement.

1.3 Energy Conversion in Batteries

A battery is a device that converts the chemical energy of the electrochemically active material on the electrode, the "active mass," directly into electric energy as represented with the solid arrow at the bottom of the diagram in Fig. 1.3.

This conversion does not involve thermal cycles, as in common combustion or heat engines where chemical energy is first converted to thermal, then to mechanical energy and finally to electrical. This is shown with the dashed arrow path in the

Fig. 1.3 Energy conversion in batteries and thermal engines

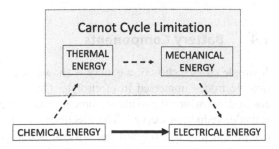

Fig. 1.4 Redox reactions on
electrodes of lead–acid battery
(left) and a
non-electrochemical redox
reaction of combustion

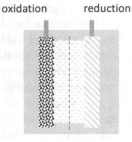

oxidation   reduction

$$0 \quad 0 \quad +4\,-4$$
$$C + O_2 \rightarrow CO_2 + \text{heat}$$

Redox reaction –
not electrochemical

Ox.: $Pb \rightarrow PbSO_4$

Red.: $PbO_2 \rightarrow PbSO_4$

diagram. These types of engines are controlled by the limitations of the Carnot
Cycle, as dictated by the second Law of Thermodynamics. In contrast, the
efficiencies of electrochemical devices, such as batteries and fuel cells, do not
have those limitations and batteries are capable of having higher energy conversion
efficiencies. That does not mean that they are 100% efficient, however, since there
are other factors affecting the reactions.

As we look more closely, the reactions involved in batteries are electrochemical
in nature, which means that the electrons are transferred from one material to another
or that the charge is exchanged between the electrolyte and the solid electrode. In
electrochemical reactions, a chemical reaction results in the production of electricity
(as in batteries or fuel cells) or electricity causes chemical reactions (as in the
electrolysis of water). An electrochemical process or device always involves a
redox reaction, which means that oxidation takes place on one electrode and
reduction on the other. This is illustrated in Fig. 1.4 on the left, using the example
of reactions in the lead–acid battery.

It should be noted that not all redox reactions are electrochemical in nature, but all
electrochemical reactions are redox reactions. To illustrate this point, we consider
combustion or burning. In this reaction, which is non-electrochemical redox reaction, the transfer of charge occurs directly, without an electrolyte and only heat is
involved. Combustion of carbon is a redox reaction and the oxidation states of
elements in the reaction are shown above the equation (Fig. 1.4, right). However,
this is not an electrochemical reaction, since there is no ionic charge transfer in
electrolyte and there is no separation of the oxidation and reduction sites.

1.4   Battery Components

A simple cross section of a battery is shown in Fig. 1.5. Batteries are comprised of
two electrodes immersed in electrolyte. One electrode is the anode, which is the
electrode on which the oxidation process takes place. On the opposite electrode, the
cathode, reduction occurs. The electrodes in batteries denote a more complex
component that usually, but not always, is comprised of a substrate or a current

Fig. 1.5 Cross section of a
battery

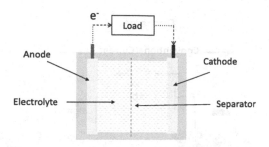

collector and active mass. In typical electrode manufacturing, active mass is coated
onto a thin metallic current collector.

In most batteries, a separator, as shown in Fig. 1.5 with a dashed line, is installed
between the anode and the cathode to prevent an electrical short. Separators are
typically an inert, nonconductive polymer material that does not prevent electrolyte
exchange and ion transport. The electrodes, together with the separator, are
immersed in the electrolyte. Electrolyte is ionically conductive medium containing
ions that carry electrical charge. Those are the basic components of all batteries.

The term battery is normally considered to be a product sold and delivered to the
final user; it is often comprised of multiple single cells, but sometimes a battery can
also signify a single cell. All this can produce a great deal of confusion. In this book,
we will refer to a battery as the end product, but we will apply an additional
description for multicell batteries using the term "stack." In most cases, when we
discuss battery chemistry, we will use the term "cell" or "single cell." Most electrical
data refer to single cells. The term "battery" will be also used to discuss general
performance characteristics.

1.5   Principle of Operation

During its operation, a battery undergoes discharge and charge processes (Fig. 1.6).
The illustration of a lithium–ion battery depicts the movement of electrons from one
electrode to another and the locations of oxidation and reduction processes.

The electrode depiction in this diagram exemplifies a typical layered structure of
the base or support electrode material, as in lithium–ion batteries, some nickel–metal
hydride batteries, as well as some primary batteries. The separator is shown in
between the electrodes. The battery is depicted as being connected to a load (left)
and a power supply (right). It is important to note the movement of electrons and
ions, the location of oxidation and reduction reactions, and the polarities of
electrodes.

Looking at the discharge reaction first (on the left), the anode is on the left, which
is the negatively charged electrode, while the cathode is on the right and it is
positively charged. As the reaction proceeds, electrons are produced in the oxidation
reaction. They flow from the location of the higher electron pressure or the negative
electrode, through the load, powering a device, and enter the cathode. During the

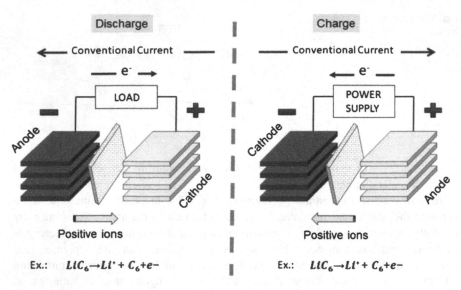

Fig. 1.6 Illustration of battery charge (right) and discharge (left) reactions

same time, the positively charged ions, lithium plus ions (Li^+) in this case, move through the electrolyte, then through the separator, and reach the cathode at the interface with electrolyte.

The reaction reverses during the charge process, which is accomplished by using a power supply or a charger. Note that the polarities of battery electrodes remain the same, the electrode on the left is still negatively charged and one on the right is still positively charged. The question now becomes how is this possible. The electrode on the left, the original anode, still has higher electron pressure or more negative potential, although the voltage difference between the two electrodes has been reduced during the discharge. For example, in the case of lithium–ion batteries, the voltage may start from 3.7 V at the beginning of the discharge and finish at 2.4 V at the end of discharge.

Normally, electrons encounter opposition flowing into the electrode on the left, which is more negative, but the power supply or charger overcomes the potential difference. It first creates oxidation and takes the electrons from the right-side electrode, then forces electrons into the more negative electrode on the left side. At the same time, positively charged ions carry ionic charge through the electrolyte from the more positive, right electrode to the more negative, left electrode. In the example of lithium–ion batteries, lithium plus ions travel from the electrode that is normally referred to as the cathode, which, in one configuration, is made of lithium cobalt oxide, through the electrolyte to reach the lithium electrode, commonly graphite, at its interface with the electrolyte.

1.6 Electrode Selection

One of the most puzzling aspects of battery design for those new to the field is the selection of electrode materials and the electrolyte. There are several important guiding principles. In the most general sense, the electrode pair is selected by looking for the largest potential difference, which would give the highest cell voltage. A simple guide, shown in Fig. 1.7 below, demonstrates how those combinations can be theoretically conceived from the table of standard reduction potentials. Note that only metals are shown in this table while other materials, such as metal oxides, require more specific information.

The direction of electron flow in a spontaneous reaction can be instantly determined from the diagram and voltage difference is calculated as the difference between the two standard reduction potentials. It is important to note that this refers to potentials under standard conditions, which means unit concentration or pressure, and at a fixed temperature of 25 °C. The actual cell voltage will be more closely predicted using the Nernst equation (see Sect. 1.8), which considers the concentration dependence of a reaction.

The second criteria when selecting electrode or active mass material has to do with the atomic or molecular weight. The lighter the material, the better characteristics a battery has in kWh/kg, which is called specific energy and is an

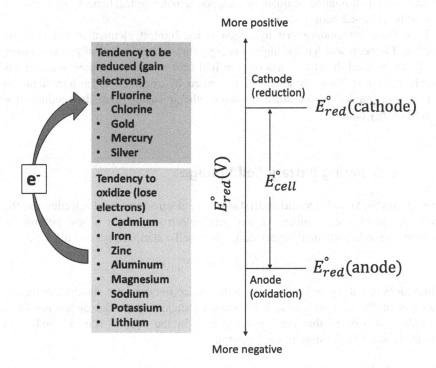

Fig. 1.7 The fundamental principle of using the table of standard reaction potentials in selecting battery electrodes

important battery characteristic for automotive and portable applications. It can be easily concluded that lithium is one of the most attractive active material candidates because it has highly negative standard reduction potential (-3.05 V) and is also very light (only 7 g/mol). It should come as no surprise that batteries based on lithium are extremely popular as of late and show a trend to replace other batteries in many applications. However, anode material, such as lithium, has to be paired with cathode material (terminology referring to the process of discharge) that is also as light as possible and has as positive standard reduction potential as possible.

Even the theoretically most promising metal–metal oxide combinations may not be practical for many different reasons, such as reactivity, slow kinetics of the reaction, interaction with electrolyte, difficulty in handling or preparation, cost, and many others. In general, anode material should have good reducing properties, while cathode material should have good oxidizing properties. Both should have high-energy content per weight and good conductivity; both should also be stable in electrolyte, be available at a low cost, and easily manufacturable.

The most common anode materials are metals such as lithium, zinc, cadmium, and others. The most common cathode materials are metal oxides; halogens, such as fluorine, chlorine, bromine, and iodine; oxyhalides, which are compounds containing a halogen atom, oxygen atom, and one more element; and sulfur. Some special battery systems can also use gaseous oxygen (from air) as the reactant on the cathode with no limitation on quantity as opposed to the typical battery. This concept is used in metal–air batteries.

It is worth mentioning that hydrogen is the lightest element in the Periodic Table of Elements and has the highest energy content per weight of any substance, but it can be used directly as gas only in fuel cells where continuous reactant gas supply is possible. There are some batteries where hydrogen is used as a reactant, but it is always bonded or contained by some other means, which then reduces the overall energy content.

1.7   Calculating Battery Cell Voltage

Armed with the knowledge and understanding of electrode material selection and the ability to calculate cell voltage for any given electrode pair, it is now possible to calculate some hypothetical battery cell voltages. To start, always use the formula:

$$E^\circ_{cell} = E^\circ_{red}(\text{cathode}) - E^\circ_{red}(\text{anode}) \tag{1.1}$$

which depicts battery cell voltage as a difference between the standard reduction potential of the cathode minus the standard reduction potential of the anode. It should be also noted that there is no change in the sign of standard reduction potentials when performing this calculation.

Example 1 The first challenge is to identify a battery electrode or reaction pair that offers the highest possible voltage of almost 6 V. This is a theoretical exercise and practicality of such a system has not been evaluated. What electrode or reaction pair can theoretically produce almost 6 V-battery?

$$\text{Li-F} : E^\circ = 2.87 \text{ V} - (-3.04 \text{ V}) = 5.91 \text{ V}$$

$$\text{Ox.} : 2\text{Li} \rightarrow 2\text{Li}^+ + 2\text{e}^-$$

$$\text{Red} : \text{F}_2 + 2\text{e}^- \rightarrow 2\text{F}^-$$

Example 2 The second question analyzes a battery composed of zinc and chlorine. What is the standard battery cell voltage for this battery? Write down the half-cell reactions while working on this problem.

$$\text{Zn}_{(s)} + \text{Cl}_{2(g)} \rightarrow \text{ZnCl}_{2(aq)}?$$

Half-cell reactions are:

$$\text{Zn}_{(s)} \rightarrow \text{Zn}^{2+}{}_{(aq)} + 2\text{e}^- \quad E^\circ_{red} = -0.76 \text{ V}$$

$$\text{Cl}_{2(g)} + 2\text{e}^- \rightarrow 2\text{Cl}^-{}_{(aq)} \quad E^\circ_{red} = +1.36 \text{ V}$$

$$E^\circ_{cell} = E^\circ_{red}(\text{cathode}) - E^\circ_{red}(\text{anode}) = 1.36 \text{ V} - (-0.76 \text{ V}) = +2.21 \text{ V}$$

Example 3 What is the standard reduction potential for the NiOOH electrode, if the standard reduction potential E^0 for Cd is -0.40 V and cell voltage for a Ni–Cd battery is 1.2 V?

$$E^\circ_{cell} = E^\circ_{red(cathode)} - E^\circ_{red(anode)}$$

$$E^\circ_{red(cathode)} = E^\circ_{cell} + E^\circ_{red(anode)} = 1.2 \text{ V} + (-0.40 \text{ V}) = 0.8 \text{ V}$$

1.8 Battery Cell Voltage and Nernst Equation

In Chap. 1, we discussed the application and importance of the concentration of reactants and products on the formation of the voltage. The electrochemical potentials and battery cell voltages discussed so far were under standard conditions: unit activity, which means that the effective concentration of a solution is 1 M or the effective pressure of a gas (known as fugacity) is 1 atm and the temperature is 25 °C. These conditions are obviously unrealistic for practical purposes in batteries and serve simply as a reference point, i.e., conditions under which the potentials of various reactions of ion electrodes in solutions can be compared. Using the standard reduction potential concept as a starting point, the actual, practical battery voltages can be predicted more accurately using the Nernst equation (see Fig. 1.8), which considers the effect of concentrations in solution or effective pressure, and temperature.

The Nernst equation connects the standard voltage of a battery and the reaction quotient for battery reactions. The voltage of a battery is, according to the equation, expected to decrease from the standard value by a factor that reflects the reaction quotient, which is the ratio of concentration of products over concentration of reactants. The natural logarithm of the reaction quotient is converted to voltage units using the RT/nF factor, where R is the universal gas constant that relates units of energy to temperature; T is the absolute temperature; and F is the Faraday constant representing the amount of electrical charge per 1 mol of electrons.

Figure 1.8 shows an example of voltage determination for lead–acid batteries (below).

In order to enter the reaction quotient, it is necessary to have a correctly balanced reaction equation for a battery. The reaction quotient expression includes the activity of 1 for all solid substances, therefore lead, lead dioxide, and lead sulfate do not appear in this equation. Also note the change in sign, from negative to positive, between the standard cell voltage and the reaction quotient term on the right. This change in sign is sometimes used when the reaction quotient is reversed, which means that the concentrations of reactants are in the numerator and concentrations of products are in the denominator.

The observation of the Nernst Equation for lead–acid batteries reveals that the cell voltage is higher than the standard battery voltage when the concentration of sulfuric acid is high, which happens at the beginning of the discharge process. As the discharge proceeds, sulfuric acid from the left side of the equation is used up to form lead sulfate and its concentration drops, lowering the reaction quotient and overall battery voltage. This is consistent with the fact that sulfuric acid concentration decreases during discharge.

In summary, the Nernst equation helps in more accurately predicting the actual battery voltage.

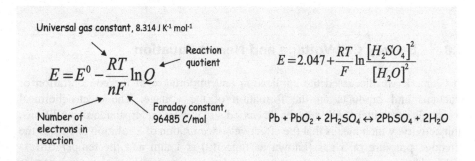

$$\text{Universal gas constant, } 8.314 \text{ J K}^{-1} \text{mol}^{-1}$$

$$E = E^0 - \frac{RT}{nF} \ln Q \qquad \qquad E = 2.047 + \frac{RT}{F} \ln \frac{[H_2SO_4]^2}{[H_2O]^2}$$

Reaction quotient

Number of electrons in reaction

Faraday constant, 96485 C/mol

$$Pb + PbO_2 + 2H_2SO_4 \leftrightarrow 2PbSO_4 + 2H_2O$$

Fig. 1.8 Schematic of the Nernst equation (left) and application to lead–acid battery voltage determination (right)

1.9   Electrolyte for Batteries

Electrolyte is a critical component for every electrochemical device, including batteries. The function of electrolyte is to ionically conduct charge between electrodes inside the cell and thereby balance the electronic charge transfer through the external circuit. The charge carried by the ions in electrolyte corresponds to the charge carried by electrons in the solid electrode. The exchange of charge takes place at the interface between the solid electrode and the electrolyte. The rate or speed of that reaction step is one of the most important factors for the overall reaction rate in any electrochemical system, including batteries.

The term electrolyte more correctly refers to a solution of electrolyte in solvent. Electrolyte is a substance that dissociates when dissolved in solvent and generates ions that produce ionic conductivity. The requirements for electrolyte are good ionic conductivity and very low electronic conductivity, no reaction with electrode materials, no change in properties with change in temperature, safety, and low cost.

Most batteries are aqueous batteries because water is the solvent used. They are formed when acid, base, or salt are dissolved in water. Their common characteristic is that they contain dissociated substances that form ions, such as H^+, OH^-, Na^+, etc. As seen in Table 1.1, aqueous electrolytes have the best conductivities of all different types of electrolytes. However, there are batteries that use organic, polymer, and solid electrolytes. In lithium–ion batteries, for example, the solvent is non-aqueous, to avoid a reaction of the anode with the electrolyte.

In some batteries, the electrolyte ions carry ionic charge and also participate in the battery reactions. One example is sulfuric acid in lead–acid batteries. The change in sulfuric acid conductivity with concentration is shown in Fig. 1.9.

In some other batteries, electrolyte conducts ions in solution, which participates in the electrode reactions, but the concentration of electrolyte does not change. An example of this is potassium hydroxide in nickel–cadmium batteries. It is worth mentioning that alkaline electrolyte is less conductive than acidic electrolytes and conductivity of 30% KOH used in nickel–cadmium batteries is approximately 0.6 S/cm. However, the alkaline electrolyte has other properties that make it a preferred choice for many batteries. The most important of these properties is the enablement of faster electrode reactions.

In lithium–ion batteries, electrolyte does not participate in the battery reactions; instead, it simply enables conduction of Li^+ ions between the electrodes. The most common electrolyte used in Li-ion batteries is lithium hexaflourophosphate. It has a

Electrolyte system	Specific conductivity, $\Omega^{-1} cm^{-1}$
Aqueous electrolytes	$1 - 5 \times 10^{-1}$
Molten salt	10^{-1}
Inorganic electrolytes	$2 \times 10^{-2} - 10^{-1}$
Organic electrolytes	$10^{-3} - 10^{-2}$
Polymer electrolytes	$10^{-7} - 10^{-3}$
Inorganic solid elec.	$10^{-8} - 10^{-5}$

Table 1.1 Ionic conductivities for most important battery electrolytes

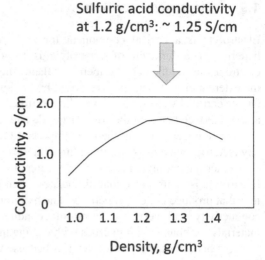

Fig. 1.9 Ionic conductivity of sulfuric acid, used as lead–acid battery electrolyte

fairly high conductivity of roughly 3 S/cm, but numerous other disadvantages, such as sensitivity to ambient moisture and solvents, thermal instability at high temperatures, and difficulty in preparation.

1.10   Gibbs-Free Energy and Battery Voltage

The most useful thermodynamic function for the evaluation of batteries is "the free energy." Free energy provides criterion for predicting the direction of battery reactions and composition of the system at equilibrium. Free energy is most often expressed using Gibbs Function and is, therefore, called the Gibbs-free energy.

The relationship between the free energy and cell voltage is one of the most important relationships in batteries. The Gibbs-free energy change equals the product of the number of moles of electrons transferred between electrodes, the electric charge per mole of electrons or the Faraday constant, and the cell voltage. The superscript "°" defines the Gibbs-free energy change and cell voltage under standard conditions of unit concentration of pressure and room temperature.

$$\Delta G^{\circ} = -n\,F\,E^{\circ} \tag{1.2}$$

Equation (1.2) provides the means to calculate the cell voltage if the electrical work or free energy is known. The negative sign consolidates the fact that the Gibbs-free energy is negative, while cell voltage is positive for spontaneous reactions, as in battery discharge.

The two examples below underscore the importance and simplicity of the relationship between the Gibbs-free energy and cell voltage for calculating the most critical parameters of batteries.

Example 4 Calculate cell voltage for alkaline primary battery with Zn and MnO_2 electrodes, if $\Delta G = -277$ kJ/mol.

The first step is to determine the number of electrons involved in the reaction by properly balancing the overall reaction.

$$2MnO_2 + Zn \rightarrow ZnO + Mn_2O_3$$

$$E = -(-2.77 \times 10^5) : (2 \times 96485) = 1.44 \text{ V}$$

Example 5 Determine the Gibbs-free energy change for the zinc–chlorine battery in acidic electrolyte, if $E^0{}_{cell}$ is 2.121 V and the number of electrons in the reaction is 2.

$$Zn_{(s)} + Cl_2(g, 1 \text{ atm}) \rightarrow ZnCl_2(aq, 1 \text{ M})$$

$$\Delta G^0 = -nFE^0{}_{cell} = -(2 \text{ mole}^- \times 96,485 \text{ C}/1 \text{ mole}^- \times 2.121 \text{ V})$$
$$= -409.3 \text{ kJ/mol}$$

1.11 Theoretical Battery Capacity

After demonstrating the calculation of the theoretical cell voltage for zinc–chlorine batteries, the next important concept for understanding a battery is the total energy it has the ability to store. It is obvious that energy depends on the amount of active material, but the relationship between the two is less clear. The link is derived from Faraday's First Law of Electrolysis, adapted to batteries, which says that the mass of a substance consumed in a battery is directly proportional to the quantity of electricity involved in the reaction.

Next, we define that 1 g-equivalent of any active battery electrode material produces 96,487 C or 26.8 Ah. Note here that the total quantity of electricity involved in the electrochemical reaction is measured in C (Coulombs) or Ah (ampere-hours). Also, 1 g-equivalent is the atomic or molecular weight of the active material in grams divided by the number of electrons involved in a reaction. Knowing the relationship between the amount of battery active mass and the quantity of electricity, it is now possible to calculate the energy contained in batteries or the battery capacity expressed in Ah. Note that the capacity in Ah, multiplied by the battery voltage gives the energy in the battery.

Example 6 Calculate the theoretical capacity of an electrochemical cell comprising Zn and Cl_2.

The molecular weight of zinc is 65.4 and the number of electrons in reactions involving zinc is 2, which gives us the equivalent weight of 32.7 g. If this amount of zinc produces 26.8 Ah then 1 g produces 0.82 Ah/g. The same calculation for chlorine gives the equivalent weight of 35.5 and 0.76 Ah/g.

$$Zn \quad + \quad Cl_2 \quad \rightarrow Zncl_2$$
$$(0.82 \text{ Ah/g}) \quad (0.76 \text{ Ah/g})$$

$$1.22 \text{ g/Ah} + 1.32 \text{ g/Ah} = 2.54 \text{ g/Ah} \quad \text{or} \quad 0.394 \text{ Ah/g}$$

$$\text{Specific Energy (Wh/g)} = 2.12 \text{ V} \times 0.394 \text{ Ah/g}$$
$$= 0.835 \text{ Wh/g} \quad \text{or} \quad 835 \text{ Wh/kg}$$

Since we want to find how much zinc chloride is needed, we can first find from the molecular weight, using simple stoichiometric calculations, that in 1 g of zinc chloride there is 0.48 g of zinc. Next, using a simple proportional analysis, we can say that if 1 g of zinc produces 0.82 Ah then 0.48 g of zinc produces 0.394 Ah. This means that 1 g of zinc chloride is equivalent to 0.394 Ah. It can be observed that it might be also useful to express the amount of material needed per Ah.

When the capacity in Ah is multiplied by the battery voltage, the energy of the battery is obtained: Watt-hour (Wh) = voltage (V) × ampere-hour (Ah). Furthermore, the energy output of a cell or battery is usually given as the ratio to its weight or volume. In terms of units, when the quantity of Ah/g is multiplied by the battery voltage, another quantity is obtained with the units of Wh/g, because $1 \text{ V} \times 1 \text{ A} = 1 \text{ W}$. This new quantity is called "specific energy" of a battery, expressed in Wh/g or kWh/kg. On a volume basis, the ratio of energy to volume of a battery is usually referred to as "energy density,, with the units of Wh/L. In certain technical environments, the term "energy density" is used for either ratio of the 2.

We see this in the example of zinc chlorine batteries where specific energy is obtained by multiplying the cell voltage of 2.12 V by capacity in Ah/g. The result tells us that this battery has the theoretical energy content of 835 Wh/kg. The same method can be used to calculate specific energy for other battery combinations.

Example 7 Calculate the theoretical specific energy for a lead–acid battery.

To determine the theoretical specific energy for a lead–acid battery, the theoretical cell voltage is first calculated from the standard reduction potentials and determined to be 2.05 V.

$$E^{\circ}_{cell} = E^{\circ}_{red}(\text{cathode}) - E^{\circ}_{red}(\text{anode}) = 1.69 \text{ V} - (-0.36) = 2.05 \text{ V}$$

Next, using the atomic weight for Pb and molecular weight for PbO_2, the electrochemical equivalents are calculated. Subsequently, the specific capacity of the product of the discharge reaction, which is lead sulfate ($PbSO_4$), is determined to be 0.12 Ah/g.

$$Pb \qquad\qquad + PbO_2 + 2H_2SO_4 \leftrightarrow 2PbSO_4 + 2H_2O$$
$$(0.26 \text{ Ah/g}) \quad (0.22 \text{ Ah/g})$$
$$(3.85 \text{ g/Ah}) \quad (4.55 \text{ g/Ah}) \qquad\qquad 8.40 \text{ g/Ah} \quad \text{or} \quad 0.12 \text{ Ah/g}$$

Table 1.2 Electrochemical equivalents for selected elements

Element	Equivalent weight, g	Standard reduction potential, V	Number of electrons, #	Equivalent capacity, Ah/g
Cd	112	−0.40	2	0.48
Zn	65	−0.76	2	0.82
Li	7	−3.05	1	3.86

After multiplying the specific capacity with the theoretical cell voltage, the theoretical specific energy of 244 Wh/kg is obtained for the lead–acid battery reaction.

$$2.05 \text{ V} \times 0.12 \text{ Ah/g} = 0.244 \text{ Wh/g} \quad \text{or} \quad 244 \text{ Wh/kg}$$

It is important to understand that the practical specific energy for lead–acid batteries is only about 35 Wh/kg (see next section for explanation).

In a similar fashion, the electrochemical equivalents for other elements, such as cadmium, zinc, and lithium are calculated using their atomic weights, standard reduction potentials, and number of electrons involved in reactions. As presented in Table 1.2, zinc has a higher electrochemical equivalent than cadmium and is, for that reason, a better anode; but lithium is, by far, the best anode material with the electrochemical equivalent of 3.86 Ah/kg.

1.12 Practical Energy of a Battery

The energy content (Wh) of a system determines its operating or battery run time. The energy values are related to the mass and volume of a cell and are expressed as specific energy (or gravimetric energy density) in Wh/kg or energy density (or volumetric energy density) in Wh/L. The theoretical specific energy and energy density can be calculated from the thermodynamic energy data and weight and/or volume of the active masses in a cell.

The energy (W), not to be confused with the symbol of the unit watt or the power, itself is the product of the average cell voltage (E) and the cell capacity (Q).

$$W = \int E(Q) \times \mathrm{d}Q = E_{av} \times Q \qquad (1.3)$$

The maximum or theoretical specific energy from a battery in Wh/kg depends on the type of elements used in active masses, their atomic weights, and their standard reduction potentials. The total energy stored clearly depends on the amount of active material or active mass, but practical energies are much lower and only a portion of the theoretical energy can be recovered as electricity. First of all, active masses are not the only components that go into assembling a battery. As examined previously, batteries encompass not only electrode active masses but also current collectors or electrode substrates, electrolyte, separator, terminals, seals, and a container. For

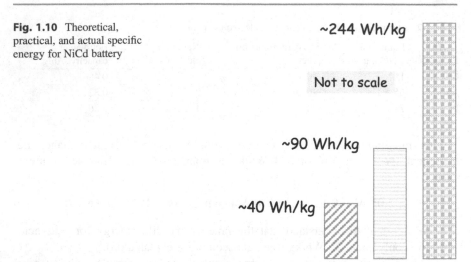

Fig. 1.10 Theoretical, practical, and actual specific energy for NiCd battery

some batteries and battery systems, additional components that add to the weight and volume include cooling and battery electrical management systems. Cell design can also affect the weight, primarily through the thickness of electrodes and separator, and create differences in the specific energy even for the same battery composition (often termed the battery chemistry).

All these additional components add to the weight and volume of a battery although they don't directly contribute to energy stored. After adding these materials, the weight of a battery increases while the stored energy is the same as obtained from the active masses. This now yields a new value for specific energy, called practical specific energy. It can be seen in Fig. 1.10, for the case of nickel–cadmium batteries that the weight of materials used for building a battery reduces the specific energy by approximately 50%.

Further reduction in the actual energy of a battery that can be converted to electricity is caused by electrochemical phenomena, such as slow reaction activation on the electrodes, incomplete utilization of active materials, ionic and electronic resistance (called ohmic losses), and mass transport losses that come from the active mass concentration reduction at the electrode. These losses typically cause a 50–75% reduction in actual energy.

After considering losses due to supporting materials and the nature of electrochemical reactions, the actual battery energy is typically 25–35% of the theoretical energy of the active materials. Theoretical and practical energy densities for some of the most important battery systems are given in Table 1.3.

As it will be described later in the text, battery capacity depends on the cell temperature and discharge current; these factors also influence the specific energy. For proper consideration of the battery specific energy values, a reference should be made to those conditions.

Table 1.3 Theoretical and practical specific energy of different battery systems

System	Theoretical specific energy (Wh kg^{-1})	Practical specific energy (Wh kg^{-1})	Percentages from theoretical (%)
Lead acid, LAB	167	33	20
Nickel–cadmium, NiCd	240	45	19
Nickel–metal hydride, NiMH	300	79	26
Nickel zinc, NiZn	320	80	25
Zinc bromium, ZnBr	435	90	21
Lithium ion, Li-ion	450	120	27
Sodium sulfur, NaS	795	90	11

1.13 Specific Energy and Power

In the previous sections, we have defined the concepts of specific energy of a battery and battery capacity. A related battery characteristic and performance indicator are the battery power, which gives an indication of the current that can be obtained from a cell. It is expressed in units of watt (W) and it becomes another characteristic value when divided by battery weight, a power density. If expressed per weight of a battery, it is usually referred to as specific power; when expressed per volume, it is called power density. However, as with energy density, power density can mean both volumetric and gravimetric power density. These two battery characteristics, energy density and power density, are the most important parameters for general assessment of battery capability and performance. They are usually presented in the form of so-called Ragone plots, where energy density is shown on the y-axis and power density on the x-axis. Thus, one graph gives instantaneous insight into the battery characteristics. It also serves as a great way to compare different batteries and, more generally, all energy storage devices. An ideal energy storage method should be shown in the upper right corner of the graph.

It would be perceptive to conclude that energy density and power density are not directly connected or dependent on each other and that this type of graph simply illustrates two battery characteristics in the same graph.

The Ragone plot in Fig. 1.11 shows specific power versus specific energy for the main battery systems, as well for fuel cells and the internal combustion engine in the upper right corner. Each energy storage system is represented by the area in the graph representing ranges of specific power and energy. The reason for this is that the information for many different batteries is presented and these ranges of values are expressed as the area on the graph.

Through further observation, it is apparent from the plot that lithium battery systems exhibit, at around 120 Wh/kg, better specific energy than nickel–cadmium, at around 40 Wh/kg, and lead–acid systems, at 30–35 Wh/kg. However, lithium–ion

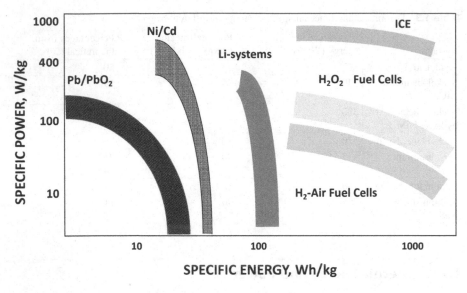

Fig. 1.11 Specific power and specific energy for selected energy storage systems

batteries have lower specific energy than fuel cells, which do not have the capability for high specific power. Nickel–cadmium batteries have the best performance for battery applications that require high power or high current.

1.14　Battery Testing

It is critically important for the study of batteries to have methods to analyze their performance. The first level of battery characterization involves performing charge–discharge cycles and measuring the capacity on each cycle. The battery capacity should ideally stay above 98% of the first cycle for several hundred cycles. Any decrease in capacity is a disadvantage for that type of battery. For capacity measurements, a charger or power supply is needed for the charge cycle and load bank for the discharge cycle. In addition, a voltmeter is needed to measure battery voltage. The battery performance is typically presented as a plot of battery voltage as a function of percentage of battery discharged, as shown in Fig. 1.12.

In addition, the temperature of a battery can be measured for safety reasons and to estimate the efficiency of converting the chemical energy of the active mass to electricity. A plot of temperature as a function of time of discharge is shown in Fig. 1.13, along with the cell voltage.

The next level of sophistication when characterizing a battery is accomplished using an instrument called a potentiostat. It enables running charge and discharge experiments using either constant current, constant voltage, or constant power. The instrument captures voltage and current during charge–discharge experiments and can display capacity and depth of discharge. In addition, the instrument has the

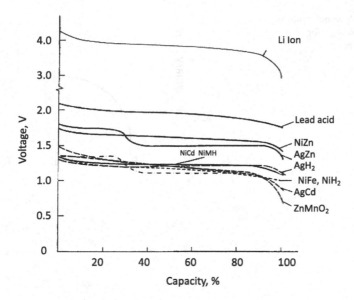

Fig. 1.12 Voltage versus % of capacity for average performance of selected commercial batteries

Fig. 1.13 Voltage and temperature versus time of discharge for a nonspecific battery

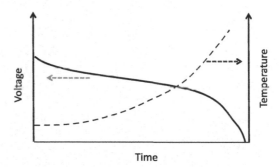

capability to conduct sophisticated measurements of battery impedance. This is done using a technique called electrochemical impedance spectroscopy. The instrument applies a small AC signal superimposed on the DC battery voltage and measures the resulting impedance over a range of frequencies. The experiment yields a Nyquist plot of the imaginary impedance versus real impedance and can produce valuable information about the performance and state-of-health of batteries. An example of a Nyquist plot is shown in Fig. 1.14.

Fig. 1.14 A Nyquist plot of battery impedance for nonspecific battery

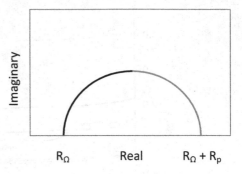

Operational Factors of Battery Systems

2

The battery performance data given throughout this textbook are based on selected and average designs and conditions of use. These values are helpful in establishing average performance characteristics, but the actual performance of the battery may be significantly different under practical conditions. The performance of a battery under specific conditions should be obtained before any final comparisons are made or definitive designs are attempted.

In addition, even for the same battery type, chemistry, or design, there are sometimes significant performance differences based on the manufacturer or between different versions of the same battery. For the same manufacturer, there are also performance variables in the production from lot to lot. The extent of variability depends on the process control specific to each manufacturer.

Throughout the text in this section, conceptual battery performance and trends are presented, without reference to any particular battery, experiment, or manufacturer. These conceptual diagrams and data serve to educate about the trends, principles, and average values; they are not referenced to any previously reported specific data.

2.1   Performance Parameters

The main operational parameters of batteries include cell voltage (V), capacity (Ah), energy (Wh), energy density (Wh/L), specific energy (Wh/kg), state of charge (SoC/%), depth of discharge (DoD/%), voltage limitation (charge, V), cut-of-voltage (discharge, V), effect of temperature, cycle life, calendar life, and internal resistance (Ω).

The state of charge and depth of discharge are depicted in the diagram in Fig. 2.1, in the form of bars that represent levels of battery capacity. The shaded part of the bars is the remaining capacity or a gauge. On the left side of the diagram is the depth of discharge. The top of the bar would mean a fully charged battery or zero depth of discharge, while the bottom signifies fully discharged battery or 100% depth of

S. Petrovic, *Battery Technology Crash Course*,
https://doi.org/10.1007/978-3-030-57269-3_2

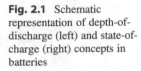

Fig. 2.1 Schematic representation of depth-of-discharge (left) and state-of-charge (right) concepts in batteries

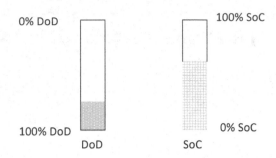

discharge. The example in the diagram indicates approximately 80% depth of discharge based on the height of the textured portion of the bar or the gauge.

The state-of-charge schematic is shown on the right and it marks the available capacity expressed on the top of the bar, which equals 100% state of charge when the battery is full, while the bottom would be 0% state of charge, in other words, the battery is fully discharged. The example in the diagram indicates about 70% state of charge. It should be obvious that the sum of the state of charge and the depth of discharge is 100%, so in the example on the right, the depth of discharge is approximately 30%. The state of charge should be always measured relative to rated capacity at the beginning of its life or use. However, sometimes the state of charge can erroneously be measured relative to the previous cycle capacity. Knowing that the capacity of a battery decreases with the number of cycles it is obvious that referencing the state of charge to the latest cycle does not give the correct information about the capacity. For many batteries, the end of life is considered to be the point when the battery capacity drops to 80% of the initial rated capacity. If the state of charge is determined relative to the last cycle it may show, for example, that the battery is 100% charged, while in fact it has only 80% of the initial capacity.

Another operational factor is the state of health, which compares the battery performance at any point with the performance of new battery. There are several factors that are included in the state of health evaluation and they are cell voltage, internal resistance, the rate of self-discharge, and the ability to accept charge. It should be clear that the state of health cannot be an absolute measurement, rather just a comparison of factors. By studying these parameters, it might be possible to get an estimate of the lifetime left in the battery. Using the metaphor from automotive technology this can be thought of as an odometer. Obviously, the state of health of a battery declines over time and number of cycles as a result of processes on materials that result in progressive deterioration until a battery cannot deliver its required performance anymore. Since state of health is a relative measurement, it is also subjective and open to interpretation relevant to specific battery, application, and other conditions. Developing meaningful set of rules or criteria is critical for maintaining a valid analysis of the state of health of a battery that can be applied for useful comparisons.

2.2 Battery Voltage

One of the most important factors for battery operation is the voltage change during charge and discharge. The voltage limitation during the charge process is shown for a hypothetical battery in Fig. 2.2. The x-axis is the time of discharge or the percentage of the capacity discharged and the shape of the curve is the same in each case, namely, the two categories or variables are equivalent and interchangeable in this context because, at constant current discharge, a certain percentage of the battery capacity is discharged in a fixed unit of time. The shape of curve in the graph is completely arbitrary, it does not represent any actual data for a battery and simply serves the purpose of illustrating the most typical general battery voltage trend during the charge process.

The graph indicates that there is a maximum voltage to which a battery should be charged. Charging a battery above that voltage is called overcharge. Most battery charging processes use constant current and, as seen in the graph, the voltage gradually increases as the battery condition is changed from discharged or partially charged to fully charged, while the battery state of charge is changing from 0% or some low percentage toward 100%.

It should be understood that the battery voltage is never zero or a low value even when the battery is fully discharged. The reason for that is that even for the conditions of complete discharge or zero state of charge there are still some active electrode materials and even the smallest quantities of these materials create difference in electrode potentials and cell voltage.

Attempting to charge a battery beyond the charge voltage limitation may result in physical damage to the battery and immediate as well as long-term negative effects. The immediate negative effects may include a sharp and dangerous temperature raise and even thermal runaway and explosion (see Sect. 2.6). This can occur when current is forced into the negative electrode despite increasing resistance and when there is no more active material that can be converted in a redox process. Since current has to be used in some way, a different reaction from the actual battery reaction—an unwanted reaction, takes place. For batteries that use aqueous electrolytes this unwanted reaction is the electrolysis of water, while for lithium batteries, the decomposition of organic electrolyte occurs.

If a battery is subject to overcharge, it is possible that the rate of heat generation exceeds the rate of heat dissipation and the result can be the thermal runway. This is

Fig. 2.2 Battery voltage during charge process

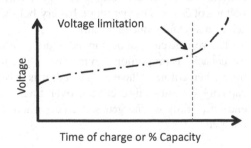

Voltage limitation

Time of charge or % Capacity

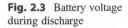

Fig. 2.3 Battery voltage
during discharge

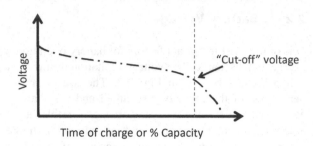

particularly serious problem for sealed batteries. The excessive heat generation process starts when I^2R losses from the current through the cell are released in the electrolyte, which lowers the electrolyte resistance and increases the current; then in turn increases the I^2R losses. This can lead to a thermal runaway and catastrophic destruction of a battery. In addition, the charging current produces exothermic reactions of the active materials and further heating of a battery. These conditions are augmented by high ambient temperature and inadequate battery design related to cooling. In lithium–ion batteries specifically, a thermal runaway can result in the meltdown of a cell or pressure increase that can lead to fire or explosion. The phenomena can also occur in valve-regulated lead–acid batteries, but not in flooded lead–acid batteries where the overcharge process causes water electrolysis and gas evolution. While not posing an immediate danger, this is still damaging to the battery cycle life.

In Fig. 2.3, a typical discharge curve showing voltage as a function of time is presented.

The curve, which is typical for most batteries, displays three regions: an immediate voltage reduction at the beginning of discharge, linear region, and sharp voltage drop near the end of discharge. For some batteries, the voltage curve in the middle region decreases less gradually and their state of charge can be estimated by the voltage value. For some other batteries, the curve is rather flat for a large portion of the discharge and estimating state of charge cannot be done using voltage value.

Another important characteristic of the curve in the graph is cutoff or final voltage, indicated by the dashed vertical line. This is the voltage at which discharge is complete and a battery should not be discharged below cutoff voltage to prevent damage. Different batteries have different cutoff voltages, for example, nickel-based batteries have a cutoff voltage of 1.0 V, while lithium–ion batteries have a cutoff voltage of 3.3 V. Discharging a battery below this voltage can lead to instability of active mass and reduced lifetime.

The voltage curve shown in the figure is typical for constant current discharge and the capacity is proportional to the discharge current. From the discharge current and time when cutoff voltage is reached, it is possible to calculate the practical battery capacity by integrating current over time, starting from fully charged state and ending at fully discharged state, i.e., when the cutoff discharge voltage (E_{cod}) is reached.

$$Q = \int_0^T I(t)\mathrm{d}t = I \times t_\mathrm{d} \tag{2.1}$$

2.3 Secondary Battery Systems

As the analysis into operational factors of batteries progresses, the text requires at this point a brief description of the major secondary battery systems that would complement the review of the operational parameters. While these systems will be studied in detail in subsequent sections, a brief analysis will compare their performance characteristics. The battery systems, listed in chronological order of their commercialization and summarized in Table 2.1 are: lead–acid, nickel–cadmium, nickel–metal hydride, and lithium–ion. The first three battery systems use aqueous electrolytes while lithium–ion uses organic electrolyte. The lead–acid batteries use an acidic electrolyte: sulfuric acid, while nickel-based systems use potassium hydroxide, an alkaline electrolyte. The most commonly used electrolyte for lithium–ion batteries is lithium phosphor hexafluoride.

A pattern of anode and cathode material pairs for various batteries should be evident. Anode is typically a zero-valent metal, lead in the case of lead–acid battery, cadmium for nickel–cadmium batteries, and lithium for lithium–ion batteries. The anode in nickel–metal hydride batteries is a special case of metal compound with hydrogen. The cathode material is a type of metal oxide, such lead dioxide in the lead–acid battery, nickel oxyhydroxide in case of nickel-based batteries, and cobalt oxide in case of the most common lithium–ion battery.

Noteworthy is also the comparison of rated single cell voltages. Nickel-based batteries have the lowest cell voltage of 1.2 V. Lead–acid battery has a rated voltage of 2.0 V, which is the highest single cell voltage of all aqueous systems. Finally, lithium–ion batteries have the highest rated voltage of all batteries of close to 4 V and while actual voltages vary for different cathode materials the most common cathode, CoO_2, has the rated voltage of 3.7 V. From the fundamental principles of batteries, it should be clear that the reason for this large battery voltage is the extreme electronegativity of lithium of -3.05 V versus the standard hydrogen electrode. This fact underscores the advantages of lithium as the anode material.

Table 2.2 shows parameters for additional alkaline battery systems.

Table 2.1 Overview of the secondary battery systems

System	Voltage	Anode	Cathode	Electrolyte
Lead–acid	2.0	Lead	PBO_2	Aq. H_2SO_4
Nickel–cadmium	1.2	Cadmium	NiOOH	Aq. KOH
Nickel–metal hydride	1.2	MH	NiOOH	Aq. KOH
Lithium ion	4.0	Li (C)	$LiCoO_2$	$LiPF_6$

Table 2.2 Additional alkaline battery systems

System	Anode	Cathode	Electrolyte
Nickel–iron (NiFe)	Fe	NiOOH	KOH
Nickel–zinc (NiZn)	Zn	NiOOH	KOH
Zinc–silver oxide (ZnAgO)	Zn	AgO	KOH
Cadmium–silver oxide (CdAgO)	Cd	AgO	KOH
Nickel–hydrogen (NiH)	H_2	NiOOH	KOH
Zinc–manganese oxide (ZnMnO$_2$)	Zn	MnO$_2$	KOH

This overview of alkaline systems suggests that many different electrode combinations were tested in alkaline electrolyte. While lead–acid battery is the only battery with the acidic electrolyte, there are many batteries using alkaline potassium hydroxide. This electrolyte, despite somewhat lower ionic conductivity than acidic sulfuric acid offers other advantages, the most important of which are fast electrochemical reactions and stability of electrode materials. Besides nickel oxyhydroxide, the other cathode materials are silver oxide, in cadmium and zinc–silver oxide batteries, and manganese dioxide in zinc manganese oxide batteries. It should also be noted here that the popularity of the manganese dioxide cathode had extended into lithium–ion systems. One of the common lithium–ion battery chemistries is Mn_2O_4 cathode material known to produce high currents and significant power.

One of the key characteristics for comparison of battery systems is the cell voltage during discharge. A plot of voltage against diminishing battery capacity, starting from fully charged battery or 100% capacity and finishing at a point when useful capacity of a battery has been exhausted and voltage fails abruptly is shown in Fig. 2.4.

The graph contains a number of practical battery cell voltages. Observation of the actual values reveals that the lithium–ion battery has the highest single cell voltage, almost 4 V, because it employs an organic electrolyte and is, therefore, not limited by the water electrolysis reaction as aqueous electrolyte systems are. It should be already clear at this point that lithium–ion battery could not operate in aqueous electrolyte because high voltage, resulting from extreme electronegativity of lithium, would cause the system to undergo water electrolysis instead of battery reaction. Also, a water-based system would not be possible due to high reactivity of lithium in water. The high single cell voltage of lithium–ion batteries means two main advantages: a clearly higher energy delivered from the battery and the simplicity to either power certain electronic devices using a single cell or a lower number of single cells in a battery stack, which in turn means smaller stack and less interconnects.

From the graph in Fig. 2.4, an aqueous battery with the highest cell voltage is the lead–acid battery at 2.05 V. The rest of the curves are for alkaline electrolyte batteries; all with less than 1.5 V and the most important systems, nickel–cadmium and nickel–metal hydride with single cell voltage of 1.2 V.

Furthermore, the shape of the voltage curves reveals that while a few alkaline batteries show voltage step, most of the curves are characterized by three regions, the

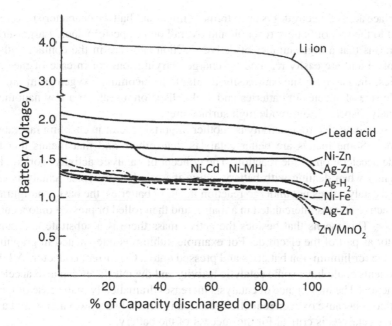

Fig. 2.4 Average cell voltage as a function of capacity for selected batteries

initial decline after start of discharge, steady voltage, and flat profile for the majority of discharge, and final abrupt voltage drop when useful battery capacity is exhausted. The flat and stable voltage is important battery characteristic as it ensures proper functioning of an electronic device that the battery is powering. The secondary batteries represented in Fig. 2.4 show flat voltage profiles for the main part of discharge, but this is not always the case and some other batteries have voltage profiles that are much steeper. In some cases, the useful capacity of a battery cannot be fully utilized because voltage decline below the value required for the operation of electronic device makes a battery impractical.

Besides energy density and cell voltage, there are additional operational characteristics used to compare battery systems and some of them will be discussed later in the text. These include cycle life, self-discharge, charge time, overcharge tolerance, deep-discharge tolerance, operating temperature, environmental impact, and cost.

2.4   Battery-Limiting Factors

The main operational attributes of batteries depend on mechanistic and structural limitations: surface area of the electrodes, catalytic activity, electrolyte conductivity, and nature of reactions on electrodes. These are some of the properties of battery components that determine battery performance or output capability.

Surface area of electrodes is an extremely important battery characteristic directly related to the rate of battery reaction and overall output performance. Large surface area means that a high number of active reaction sites are in the contact with the electrolyte and are easily reached by charge-carrying ions; or in case of metal–air batteries, the catalytic sites are easily available to incoming oxygen from air. The particle size of species in batteries and fuel cells is on a scale of a few nanometers and finely dispersed to provide high surface area.

The catalytic electrode activity is another important factor in enabling fast rates of reactions. Some metals are better catalysts than others and that means that they provide acceleration of the reaction. The concept of catalytic activity works a little differently when dealing with battery active masses than in typical catalysis cases where a catalyst does not undergo reaction itself. In batteries, the basic configuration is that active mass is intercalated in a matrix and then rolled or pressed onto a current collector. This means that besides the active mass there is a substrate and current collector as part of the electrode. For example, lithium is intercalated in graphite in the anode for lithium–ion batteries and pressed onto a Cu current collector. All three components could have some catalytic activity and the interfaces can also accelerate the reaction. The importance of catalysis increases dramatically in the case of metal–air batteries because oxygen reduction reaction is normally a slow reaction and using effective catalysts is critical for the success of the battery.

Diffusion of electroactive species through the electrolyte is another important factor in reaching a desired operational output of a battery. This characteristic is related to internal resistance and it clearly depends on ionic conductivity of electroactive species in the electrolyte. Using the example of lithium–ion batteries, an electroactive species is Li^+ and its diffusion through the electrolyte is one of the critical reaction steps for the overall rate of reaction.

The overall battery behavior, which determines the operation, depends on the ability to restrict reactions to just the desired battery reaction. However, in all batteries, there are side reactions that occur and don't contribute to faradaic reactions that produce power. The extent of these reactions depends on the system, combination of elements, and interaction with electrolyte. The type of side reaction is different for every battery. For example, in lithium–ion batteries, a well-studied reaction is formation of solid electrolyte interphase or SEI layer. This undesired reaction contributes to loss of active material, removal of electrolyte, and increased battery internal resistance. Minimizing side reactions by various strategies is the key to improving battery performance.

2.5   Battery Current

The current used to charge or discharge a battery is typically expressed as a multiple of a so-called C-rate, which is the capacity of a battery. For example, a 1 C current is the current expressed in amperes (A) that has equal numerical value to the battery capacity in Ah, i.e., it represents the current that would charge or discharge a battery in 1 h.

$$I(A) = M \times C \tag{2.2}$$

where M is the multiple of capacity, C.

The actual expression for this current is actually not dimensionally correct because a multiple of capacity will be in Ah and not in A, as required for current, but this way of expressing current has been accepted in the battery industry.

Example 1 What is the C-rate if a current of 300 mA is used to charge a cell with a rated capacity of 1200 mAh?

$$M = I/C = 300 \ mA/1200 \ mAh = 1/4$$

The current chosen is expressed as C/4 or 0.25 C. The battery would be fully charged using this current in 4 h.

What is the C-rate if a battery has a capacity of 1200 mAh and is discharged using 2.4 A current?

$$M = 2400 \ mA/1200 \ mAh = 2 \quad or \quad the \ C\text{-rate is } 2 \ C.$$

Example 2 What is the battery capacity if it is fully charged using C/10 current of 250 mA?

$$C = I/(C/10) = 250 \ mA \times 10 = 2500 \ mAh$$

As needs for some battery tests require constant power discharge, a method for definition of power value has been defined, which is simply a product of rated energy in Wh and a certain multiplication factor. For example, a battery rated at 5.6 Wh will be charged in 30 min if power of 2 P or 11.2 W is applied.

The current and power used to charge and discharge batteries have now been defined.

2.6 Modes of Discharge

A battery can be discharged using constant current, constant load or resistance, and constant power. Variations are also possible, especially for realistic battery applications such as automotive load regime, which is shown in Fig. 2.5 and which depends on the driving speed.

It is expected that curves for both current and voltage as a function of time will be different for the three different discharge modes. In Fig. 2.6, the current is shown for three modes.

It is clear that the current itself will be constant for the constant current discharge; this is horizontal line in the graph. For the constant power mode, current increases

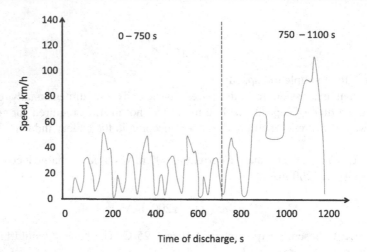

Fig. 2.5 Simulated battery driving cycle. Adapted from various resources

Fig. 2.6 Current versus time
of discharge for constant
power and constant resistance
discharge modes. The
horizontal line represents
constant current discharge

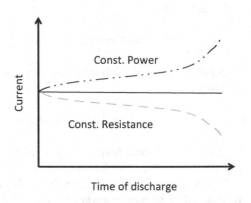

with progression of discharge because voltage decreases. For constant load conditions, which is the most realistic discharge mode for majority of applications, current decreases with voltage decrease to counter the unchanging resistance (i.e., load) conditions.

The voltage change for three modes of discharge is shown in Fig. 2.7.

In each case, the shape of curve is similar, but useful capacity is exhausted and sharp voltage drop takes place at different times. This point is reached the soonest in the case of constant power. It is followed by a longer discharge time for constant current (solid line) simply because the voltage drop at constant current means decreasing power demand with decreasing voltage. Finally, constant resistance takes the longest time to reach the full discharge because the current adjusts with the voltage drop to keep the load constant, hence both voltage and current decrease means that the discharge slows down near the end of the process.

Fig. 2.7 Voltage as a
function of time of discharge
for constant power and
constant resistance mode.
(Adapted from a variety of
sources for illustrative
purposes)

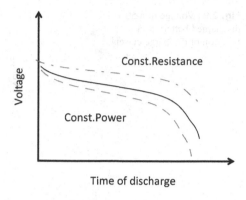

Fig. 2.8 Voltage of a battery
as a function of current

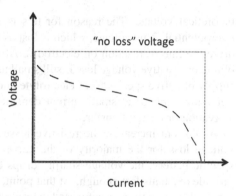

2.7   Discharge Current Effect on Voltage

The effect of discharge current on cell voltage is one of the most important relationships relative to battery operation. It tells us what a battery voltage would be if discharged with different currents and it is obvious that the higher the discharge current, the lower is the voltage. At higher currents, more electrons are being removed from the anode and transferred into the cathode and more active mass is converted on both electrodes; on the anode, the active mass is oxidized, for example, cadmium into cadmium hydroxide and on the cathode, nickel oxyhydroxide is reduced to nickel hydroxide. Since cell voltage is the potential difference between the two electrodes, the more charge transferred from anode to cathode means that this potential difference becomes smaller, hence the voltage is lower. The graph in Fig. 2.8 exhibits familiar shape of the curve featuring three regions.

The overall behavior is considerably different from the constant or "no loss" voltage, based on theoretical calculation. The theoretical assumption is that only after active mass is completely used in a reaction will the voltage drop to zero.

Practical voltage is, however, very different. In the first region, even at small current densities, the voltage is lower than the open circuit or thermodynamic, i.e.,

Fig. 2.9 Voltage of a 50%-discharged battery as a function of discharge current

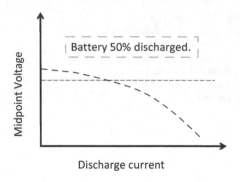

theoretical voltage. The reason for this is a voltage drop described as activation overpotential. This voltage, which is lost as useful electrical energy, is sacrificed to drive the rate of reaction on electrodes. At this point, at low current densities, the ohmic or resistive voltage loss is still small as the $I \times R$ is small and the need for fast supply of active species from electrolyte and the far regions of the electrode is still not critical, hence at small current densities the losses are due to process called activation or charge transfer.

As current increases, the resistive losses become significant and dominate the voltage loss for the majority of the curve. At the highest current densities for a specific battery, the voltage sharply drops because of the inability of a system to provide reactants fast enough. At that point, there are really no reactants left, which means that everything is converted to products. For example, in the case of a nickel–cadmium battery during discharge, all cadmium on the anode or most of it is already converted to discharge reaction product, cadmium hydroxide. So, the concentration or availability of cadmium approaches zero and causes a voltage drop, which leads to the end of discharge reaction.

A midpoint voltage is the nominal voltage of a battery and is measured when a battery is discharged to 50% of the capacity. A plot of the midpoint voltage versus current is given in Fig. 2.9.

The horizontal line represents the midpoint voltage for the nominal current. Below this nominal current, the midpoint voltage is higher and above the nominal current the midpoint voltage is lower. For current higher than the nominal, the midpoint voltage value is lower. It is important to know the midpoint voltage as a function of current to properly design and select a battery for a certain application. Each application, i.e., a device has the required minimal voltage it must obtain from the battery. The load is also defined by the device, which in turn determines the current. Using voltage versus discharge current plots, it is possible to match a battery with the application.

Figure 2.10 shows another way to look at the effect of current on the battery voltage. Three different, hypothetical discharge currents produce three different voltage curves as a function of the time of discharge. The higher the current, the

Fig. 2.10 Voltage versus time of discharge for three different hypothetical discharge currents

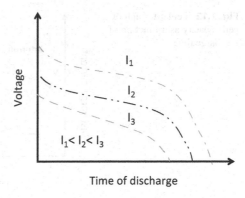

Fig. 2.11 Capacity versus discharge current. Adapted from a variety of sources as conceptual trend

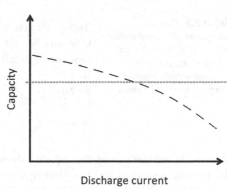

lower is the curve and the shorter the discharge time. This means that the useful capacity of a battery is smaller at higher currents.

2.8 Discharge Current Effect on Capacity

Battery capacity is the total number of ampere-hours that a battery can deliver under specified or nominal discharge current and temperature. This is a quantity declared by a manufacturer to be the output of the battery when discharged from full charge to cutoff voltage. It is easily obtained for the constant current discharge by integrating current over time (Eq. 1.3). Since nominal capacity is given for a certain discharge current, it must be that the capacity changes with the value of the discharge current (Fig. 2.11). The nominal capacity is obtained for certain, nominal discharge current and is indicated by the horizontal line. For currents lower than the nominal discharge current, capacity higher than nominal or 100% is obtained, while for currents higher than nominal the capacity is lower.

The Peukert equation or curve describes how capacity depends on the discharge current. It has been historically used for lead–acid batteries. In the graph in Fig. 2.12, the lower solid curve signifies how discharge current changes with the time of discharge. It should be clear that longer discharge time means lower discharge

Fig. 2.12 Peukert graph of cell capacity as a function of discharge time

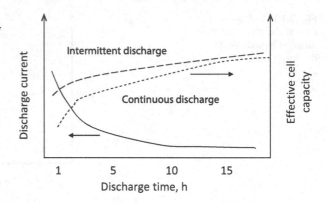

Table 2.3 Peukert coefficient (pc) for different battery systems and types for room temperature. Average values adapted from a variety of sources

Battery system/type	pc
Lead–acid, flooded	1.2–1.4
Lead–acid, VRLAB	1.15–1.25
Ni–Cd	1.10–1.20
Ni–MH	1.05–1.15
Li–ion	1.05–1.1
Ideal battery	1.0

current. At the same time, the capacity shown on the y-axis on the right increases as the current decreases and the discharge time increases. The intermittent discharge curve shows higher capacity than the continuous discharge curve.

The Peukert equation gives the means to calculate capacity for a specific current if capacity is known for another current.

$$Q_2 = Q_1 \left(\frac{I_1}{I_2} \right)^{pc-1} \tag{2.3}$$

The calculated capacity Q_2 is mainly dependent on the Peukert coefficient (pc), which is related to internal resistance of the cell and varies not only by battery systems but also by design of the same battery chemistry, temperature of operation, and age of a battery. Selected Peukert values are shown in Table 2.3.

The effect of the Peukert coefficient is shown in Fig. 2.13 for an example of hypothetical 120 Ah battery. It is evident from the figure that there is a strong dependence of the battery capacity on the Peukert coefficient and discharge current.

2.9   The Effect of Temperature on Battery Performance

The effect of temperature on cell voltage and battery capacity is one of the critical factors influencing the operation of batteries. Since battery processes are electro-chemical in nature and all chemical reactions exhibit higher rates as temperature increases, the rates of battery reactions also increase with temperature. As a result, a

Fig. 2.13 Capacity of a
hypothetical 120 Ah battery
for different Peukert
coefficients

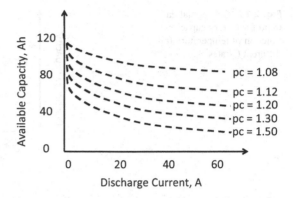

Fig. 2.14 Battery voltage as
a function of capacity for
different temperatures. The
curves in the figure do not
represent real data, only
conceptional trend

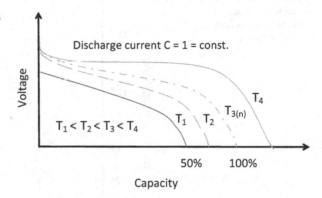

general trend is that battery operational factors such as cell voltage and capacity increase as the temperature increases. The effect of temperature is not as simple as presented here and proper evaluation would reveal a complex interaction of factors that is beyond the level of examination in this text.

A hypothetical battery record in Fig. 2.14 reveals a trend of how cell voltage is affected by the temperature. One hundred percent capacity should be obtained at the nominal temperature of 25 °C (curve 3). At higher temperatures, higher capacity is obtained and at lower temperature, lower capacity.

The effect of discharge current or the C-rate on the battery voltage has been previously discussed in Sects. 1.7 and 1.8. The two parameters, C-rate and temperature, both affect battery voltage. Lower temperatures and higher C rates result in lower cell voltages; and higher temperatures and lower C-rates lead to higher cell voltages. It can be concluded from considering these two factors that under temperatures lower than 25 °C and discharge current higher than the nominal discharge current, the nominal capacity, or 100% of capacity cannot be achieved. For example, for a low temperature and high C-rate such as 8 C, the actual capacity most batteries can deliver could be only around 30% of the nominal. An increase in C-rate from 1 to 8 C can typically lower the usable capacity by 20–30%.

Fig. 2.15 Conceptual data
trend for battery capacity as a
function of temperature for
different C-rates

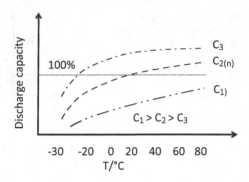

Fig. 2.16 Specific energy
versus temperature for
selected batteries from a
variety of sources

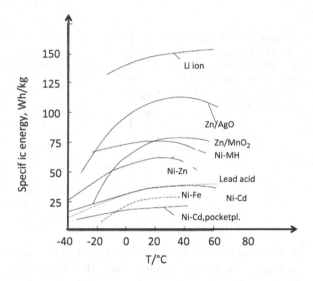

When capacity versus temperature is plotted for different C-rates a trend revealed
is an increase in capacity from lower to higher temperature and for lower C rates
(Fig. 2.15). (The curves in the graph are constructed from arbitrary data, not real
battery data.) The dashed line represents nominal capacity at a certain C-rate,
typically C/12 and nominal temperature, typically 20 °C. For higher C-rates and
lower temperatures, the capacity is lower and for lower C rates and higher temperature the capacity is higher than nominal.

The temperature also affects energy density in a similar way as it affects capacity.
Selected data for actual batteries are shown in Fig. 2.16.

Examining the real data brings the realization that the trends are only general and
that real batteries may behave differently. While most batteries in the graph exhibit
the mentioned increasing trend with temperature, some others show first an increase
followed by a peak and subsequent decrease. These are zinc-based batteries and the
nickel–metal hydride battery. As already discussed, while the primary effect of
temperature suggests an increase in operational factors including energy density,

the overall mechanistic effects are much more complex, and, at a certain temperature, additional factors influence the cell voltage, capacity, and energy density.

Example 3 An unspecified battery has nominal cell capacity for C/12 current and 20 °C. Is the cell capacity going to be lower or higher than nominal for C/10 and 25 °F? And for C/20 and 0 °C? Is it possible to give a definite answer in both cases?

The answer for the first case should be straightforward. Both C-rate (or current) and temperature trends lead to lower capacity because the C-rate is higher than nominal and temperature is lower than nominal. Hence, the capacity will be lower than nominal.

But the second case is more complicated. The C-rate is lower than nominal and should result in higher capacity but the temperature is lower than nominal and will lead to lower capacity. Therefore, these are competing trends and without knowing the particular properties of a battery or being able to collect empirical data, it would be impossible to predict if the resulting capacity will be higher or lower.

2.10 Self-Discharge

Another operational factor critical for batteries is self-discharge. It refers to a loss of capacity through involuntary reaction during the battery storage and when battery is not connected to a load. The self-discharge reactions are chemical reactions that are analogous to battery discharge reactions, but without producing electricity and powering the load. The reactions are mainly caused by thermodynamic instabilities between active masses and cell components, resulting in consumption of active masses. It should be obvious that batteries suffer loss of capacity through self-discharge reactions and minimizing self-discharge is critical for improving battery performance.

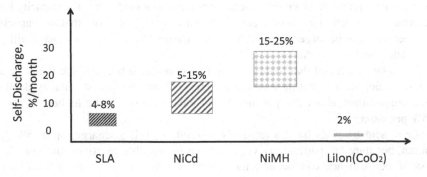

Fig. 2.17 The rate of self-discharge for four battery systems. The heights of boxes represent the ranges of self-discharge rates for different versions and designs of battery systems

Fig. 2.18 Conceptual capacity decrease as a function of time in months. No real data are used for constructing this graph

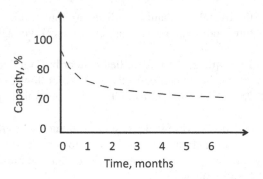

The rate of self-discharge is dependent on the battery type. Each battery system has different rate of self-discharge. Figure 2.17 compares self-discharge in percentage of capacity loss per month for the four main battery types.

It can be concluded from the graph that the lithium–ion battery has the lowest self-discharge rate while nickel–cadmium has the highest, from 20 to 25% per month. Nickel–metal hydride also has a significant rate of self-discharge per month while lead–acid features a fairly low rate of self-discharge.

The effect of self-discharge on the battery capacity is clear; the capacity decreases over time as a result of loss through self-discharge (Fig. 2.18).

Normally, the self-discharge is not an irreversible process and when a battery is recharged the capacity is restored. This is true for most batteries, but not for lead–acid batteries where the product of discharge reaction is lead sulfate, a compound that may build and grow irreversibly if the recharge doesn't occur in a certain time. Hence, it is important to recharge batteries that are not being used to restore them to 100% capacity. The self-discharge has an important consequence for battery performance—if a battery is left for a long period of time without recharging, the available capacity will be reduced.

Temperature affects battery self-discharge reactions as well. Since higher temperature promotes higher rates of all reactions it also increases the rates of self-discharge reactions. Hence, storing batteries at high temperatures is detrimental to their capacity retention since self-discharge reaction rates increase. The capacity loss per year as a function of temperature is shown in Fig. 2.19. Note that the capacity loss per year can be larger than 100%, which means that a battery capacity will be completely lost in less than a year.

It can be concluded that trends among the represented batteries are the same as before, which means that lithium–ion batteries exhibit the lowest capacity decline with temperature, about 2% per month and nickel–cadmium the highest of about 25% per month.

Lead–acid batteries have a relatively low rate of self-discharge, up to 5% per month, but they are particularly vulnerable to irreversible reactions and complete loss of performance can occur after several months of inactivity. Nickel-based batteries have particularly high rates of self-discharge, but it is relatively inconsequential for them because the reaction is not irreversible, and the capacity can be

Fig. 2.19 Capacity loss per year as a function of temperature for selected batteries. Data in this graph are conceptual and not based on actual batteries

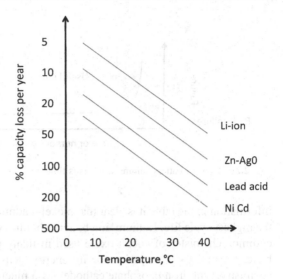

Table 2.4 Calendar and cycle life for selected battery systems

Battery	Calendar	Cycle life
Lead–acid, SLI	3–6 months	200–700
Sealed NiCd (FNC)	5–20 months	500–10,000
Nickel metal hydride	2–5 months	300–600
Nickel iron	8–25 months	2000–4000
Zinc–silver oxide	2 months	50–100
Lithium cobalt oxide	/	300–1000

restored with charging. However, the availability of capacity on demand may affect some applications.

2.11 Calendar and Cycle Life

One of the most important battery operational factors besides energy and power density is the cycle life. It determines how many times a battery will be able to deliver charge–discharge cycles without its capacity dropping below a certain level, typically 80% of the original capacity. The manufacturers specify the cycle life for certain conditions of charge and discharge such as C-rate, temperature, and the depth of discharge. Other factors may affect cycle life as well.

The cycle life is clearly different from the calendar life, which simply defines the time during which battery components age and affect the performance so that it fails below 80%. The calendar life may include time in operation or storage.

The table provides typical values for both calendar and cycle life for a number of battery systems (Table 2.4).

The cycle life for secondary batteries can vary widely, from less than a hundred to even 10,000 (cycles) for some special nickel–cadmium batteries. From the

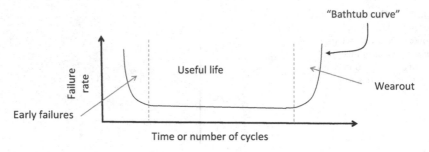

Fig. 2.20   General battery failure rate versus time

information in the table it is clear that nickel–cadmium batteries are characterized by the highest cycle life. Lithium–ion battery chemistry shown in the table has the most common chemistry of cobalt oxide used in many portable devices. This particular battery chemistry is superior for its energy density, but a different lithium–ion chemistry, with iron phosphate cathode has a much longer cycle life of up to 3000.

The problem with battery cycle life is that it can only be estimated based on general empirical information and not absolutely determined. The actual battery cycle life will be known only at the end of its lifetime when a battery fails, which means when its capacity drops below specified value. Experimental determination of the battery cycle life is obviously complicated, as it would require a very long test time and a large number of batteries would need to be tested to failure. The accelerated lifetime testing methods often used for other electronic products would not be accurate since they would involve higher C-rates, which is known to change the cycle life. Requirements for accelerated tests are that the main aging parameter is not affected by the acceleration factor (i.e., stress parameter), which is usually the elevated temperature. The interconnectedness of numerous battery design parameters and operational factors such as voltage, current, self-discharge, internal resistance, and temperature makes the lifetime modeling and prediction extremely difficult. Some methods that estimate battery state of health have been shown to provide enough insight to reasonably predict the lifetime.

In addition, batteries employed in actual applications could be operating in very different conditions of temperature and actual depth of discharge, which would make prediction of lifetime even more difficult. The general reliability of batteries can be assessed similarly to other devices and is shown in Fig. 2.20.

Failure rate is displayed versus time and there are three distinct regions in the typical shape called "bathtub" curve. Some batteries, just like other electronic devices, can have early failures due to manufacturing imperfections. After the initial period, failure rate becomes low and steady until battery materials start wearing out and reach end of life or wear out.

Besides the battery chemistry and construction, the two most critical operational factors for cycle life are depth of discharge and temperature. The depth of discharge impacts cycle life very directly for most batteries. The greater the depth of discharge, the shorter or lower the cycle life. This simply means that there is a certain battery

Fig. 2.21 Cycle life as a function of depth of discharge for a hypothetical battery at three different temperatures

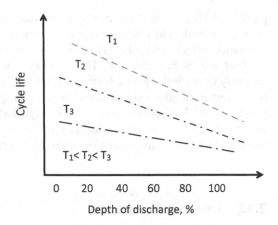

cycle life that depends on the battery chemistry, construction, and materials used; and a nominal cycle life will be determined assuming that all active electrode masses are going to be engaged in charge and discharge reactions in every cycle. But, if a battery is not fully discharged on every cycle some of the active masses are not involved in the reaction and this has a positive effect on the battery life.

Cycle life also depends on temperature. Battery lifetime or cycle life testing can be done in a laboratory environment, at a constant and controlled temperature; and most cycle life specifications are given for conditions including a nominal temperature of 20 °C. But clearly, over the lifetime of a battery the temperature will not be constant for most types of applications. This means that some average temperature will be taken into consideration when predicting the cycle life. The changes in temperature also have effects on battery components and particularly on the interfaces between the battery components. In general, the higher the temperature during battery cycling and storage, the lower the cycle life. As discussed before, higher temperature contributes to higher rates of reaction, but also faster aging and degradation of materials. The unwanted side reactions are also faster at higher temperature. All these factors contribute to decrease in cycle life with increased temperature.

The graph in Fig. 2.21 shows general trends for the cycle life as a function of the depth of discharge for three temperatures.

The graph underscores the previously discussed trends where cycle life decreases at higher depth of discharge and higher temperature. It is also obvious from the graph, by observing the first part of the graph, that the cycle life can be much higher than for shallow cycling or for cycling that involves low depth of discharge. Effectively, a battery can last longer, in some cases much longer, if it is only partially discharged on every cycle.

However, there are problems with shallow cycling with nickel-based batteries, especially nickel–cadmium batteries. During repeated shallow cycling some of the active mass of cadmium and to a smaller extent nickel oxyhydroxide becomes inactive and that causes growth in crystal, converting active mass from small fine particles with large surface area to larger lumps with lower surface area that are difficult to break. The reduction in surface area results in capacity loss. This is called a "memory effect," as if a battery electrode remembers that it doesn't need to keep all parts of the electrode active. The memory effect can be simply prevented by regularly fully discharging a battery. This discharge cycle can take place even only once every 3 months and that would prevent the growth of cadmium crystals while preserving finely dispersed particles with high surface area. Therefore, the memory effect is not an irrepressible fundamental flaw of nickel–cadmium batteries, it is a myth based on poor understanding of the basic principles and battery materials.

2.12 Internal Resistance

The internal battery resistance is a critical property of a battery that determines how much current can pass through a cell; the lower the internal resistance the more current can flow through a battery during charge or discharge. The internal resistance is one of the best indicators of battery health. It deteriorates gradually during the life of a battery and it sometimes shows early signs of reactions that contribute to its increase.

The equivalent circuit representing a typical battery is shown in Fig. 2.22.

Activation resistance and capacitance come from processes directly on the electrode–electrolyte interfaces, while the resistance component on the right comes primarily from the electrolyte and contacts. Over the lifetime of a battery, additional resistive and capacitive components may develop as a result of processes that produce passive layers or deposits. The appearance of those additional battery interfaces usually increases the internal resistance, reduces the rate of electrochemical reaction, increases resistance and ohmic losses, and slows down diffusion of species to electrode active sites.

During discharge of a battery, the internal resistance increases as shown in Fig. 2.23. This is mainly because of the changes in the electrolyte as in lead–acid batteries or as a result of formation of passivating layers through which the reactants have to diffuse.

As temperature increases, the internal resistance decreases (Fig. 2.24). This is consistent with the previous conclusions and the fact that battery capacity improves

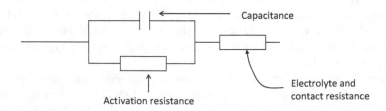

Fig. 2.22 Equivalent circuit of a typical battery

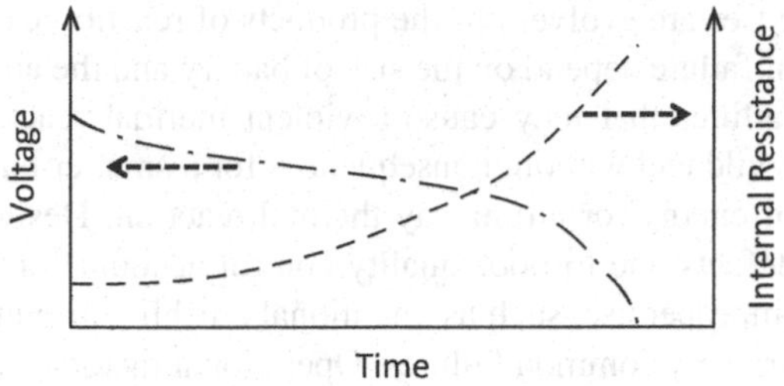

Fig. 2.23 Conceptual effect of time of discharge on voltage and internal resistance of a battery

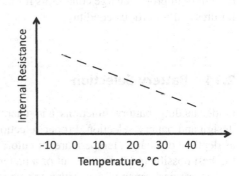

Fig. 2.24 Internal resistance trend as a function of temperature for generalized battery

with temperature. It is also obvious that ionic electrolyte conductivity increases with temperature increase due to higher ionic mobility.

2.13 Safety

Battery safety is a general term describing the ability of a battery to operate under various conditions without failure that can result in hazardous conditions or a release of toxic materials. Hazardous conditions usually refer to evolution of heat, fire, and explosion. Toxic materials can be battery electrolytes, byproduct chemicals of the unwanted side reaction, or gasses that evolve from reactions at high temperatures. Hazardous conditions and release of toxic chemicals often happen in the same battery failure event.

In general, battery safety is determined by the type of battery and electrolyte, nature and reactivity of electrode active masses, types of reactions taking place in the overcharge and over discharge conditions, overall energy content, design of electrodes, type of packaging, battery use, operational parameters, and temperature conditions. The type of battery and electrolyte it uses defines essential predisposition of a battery system for any unsafe release of energy. Systems with aqueous electrolytes are generally more stable and less prone to intense and unsafe reactions than Li–ion batteries, which use organic solvents and electrolyte. The reactivity of active masses is a critical property and those batteries with reactive active masses are more prone to failures that include gas evolution, decomposition, and such. Most catastrophic battery failures occur due to improper charging or discharging, when gases are evolved, or the products of reaction cause electrical shorts. Consequences of failure depend on the size of battery and the amount of active masses. The types of failures that may cause a violent thermal reaction and disintegration of a battery could end without consequences for a smaller battery because of insufficient release of energy for a runaway thermal reaction. Design of electrodes and manufacturing defects due to poor quality control account for most of the safety battery failures. Improper use, such as intentional crushing or puncturing, or external short circuiting are very common failures. Operational factors such as too high a charge or discharge current or improper charge conditions for the temperature lead to major catastrophic failures and hazardous conditions.

2.14   Battery Selection

Understanding battery functions and characteristics is critical for proper system design and battery selection. Proper selection of battery operating conditions, such as depth of discharge, temperature, or cutoff voltage is equally important for getting the best possible performance out of a battery.

Battery performance data given throughout this text are based on selected and average designs and conditions of use (Table 2.5). These values are helpful in establishing average performance characteristics, but the actual performance of a battery may be significantly different under actual conditions, load, temperature, state of charge, etc. The performance of a battery under specific conditions should be obtained before any final comparisons are made or definitive designs are attempted. In addition, even for the same battery type, chemistry, or battery design, there are sometimes significant performance differences based on the manufacturer or between different versions of a same battery. For the same manufacturer, there are also performance variables in production from lot to lot. The extent of variability

Table 2.5 Most important, average battery parameters for four battery systems

Parameter	NiCd	NiMH	SLA	Li–ion
Energy density (Wh/kg)	40–60	60–80	30	80–160
Cycle life (end of life at 80% capacity)	1500	1000	400	300–3000
Efficiency (%)	70	70	80	93
Optimum charge time (h)	1.5	2–4	8–16	2–4
Overcharge tolerance	Moderate	Low	High	Very high
Deep discharge tolerance	Moderate	Moderate	Low	Low
Self-discharge per month (25 °C)	20%	30%	3–5%	2–3%
Cell voltage (nominal)	1.2 V	1.2 V	2 V	3.7 V
Environmental impact	Recycle Cd	Cd-free	Recycle Pb	No heavy metals
Relative battery cost	1.1–1.2	1.4–1.6	1	3–6

depends on process control specific to each manufacturer. Particular manufacturer's data for a battery to be used should be obtained for critical designs or evaluations.

The importance of these parameters is relative and depends largely on the application. For electromobility, the most important parameters are specific energy and power; for cars with smaller available space energy and power density. For stationary applications, the most important is cycle life, followed by efficiency, self-discharge rate, and cost. For portable applications, energy density dominates among the consideration criteria because of the generally small volume requirement for a battery. Cost is more important for larger batteries than for small batteries for portable applications.

Lead–Acid Batteries

3

3.1 Overview and Characteristics

Lead–acid battery (LAB) is the oldest type of battery in consumer use. Despite comparatively low performance in terms of energy density, this is still the dominant battery in terms of cumulative energy delivered in all applications.

The working principle of LAB was discovered in 1859 by Wilhelm Joseph Sinsteden (1803–1891). The first practical battery was built by Gaston Planté (1834–1889), who developed cylindric and planar cells with electrodes (called plates) separated by rubber strips. The active masses of the negative and positive electrodes were electrochemically prepared on lead plates, a process still used even today.

Lead–acid batteries are comprised of a lead-dioxide cathode, a sponge metallic lead anode, and a sulfuric acid solution electrolyte. The widespread applications of lead–acid batteries include, among others, the traction, starting, lighting, and ignition in vehicles, called SLI batteries and stationary batteries for uninterruptable power supplies and PV systems.

From the original, flooded-type lead–acid batteries several other configurations emerged. The flooded configuration means that the electrodes are immersed in electrolyte, which is sulfuric acid, and the cells of a battery are open to air through a small vent in the cap. If such battery was opened or punctured, there would be a free liquid electrolyte spill, which makes flooded lead–acid batteries hazardous because of the significant content of liquid corrosive acid. The other emerging configurations include sealed lead–acid, gelled electrolyte, invented in 1957 by Otto Jache, and Absorbed Glass Mat (AGM), patented by Gates Rubber Corporation in 1972.

Lead–acid batteries have the highest cell voltage of all aqueous electrolyte batteries, 2.0 V and their state of charge can be determined by measuring the voltage. These batteries are inexpensive and simple to manufacture. They have a low self-discharge rate and good high-rate performance (i.e., they are capable of high discharge currents). Lead–acid batteries are mature, reliable, and a well-understood technology. When used correctly, they are durable and provide dependable service. They are available in large quantities and a variety of sizes: from 1 Ah to several thousand Ah and their electrical efficiency is higher than 70%. Based on their robustness, predictable performance, and low-cost, LABs are still the most commonly used battery system.

However, lead–acid batteries have inferior performance compared to other secondary battery systems based on specific energy (only up to 30 Wh/kg), cycle life, and temperature performance. The low-energy density limits the use of lead–acid batteries to stationary and wheeled (SLI) applications. They are prone to sulfation of the electrode plates, a process by which a product of discharge reaction on both electrodes, lead sulfate, grows in particle size, reducing the active surface area for reaction. These reactions become irreversible after a certain period of time, leading to permanent battery damage.

Another operational limitation of lead–acid batteries is that they cannot be stored in discharged conditions and their cell voltage should never drop below the assigned cutoff value to prevent plate sulfation and battery damage. Lead–acid batteries allow only a limited number of full discharge cycles (50–500). Still, cycle life is higher for lower values of depth of discharge and these batteries are well suited for standby applications that require only occasional deep discharges.

Because of the lead toxicity and corrosivity of electrolyte, the LA battery is environmentally unfriendly. The corrosivity of sulfuric acid and environmental concerns regarding spillage mean that there are transportation restrictions on flooded lead–acid batteries. With sealed lead–acid batteries, the problems of free liquid electrolyte are replaced with issues involving gas evolution and temperature rise during charging, which can lead to thermal runaway.

3.2   Principle of Operation

In the discharge reaction in the diagram (Fig. 3.1), the electrons move from left to right through an external circuit, powering the load. On the left side is the negative, lead electrode and oxidation occurs on this electrode during discharge. Elemental lead, Pb reacts with sulfuric acid during the discharge process to form lead sulfate on the electrode, while protons go in the solution and electrons exit the electrode and travel through the external circuit. The right-side electrode is the lead dioxide, PbO_2. During discharge, PbO_2 reacts with sulfuric acid, protons, H^+ from the solution and electrons arriving from the external circuit, to form lead sulfate, $PbSO_4$, and water.

Sulfuric acid participates in the reaction and it is consumed during discharge, effectively lowering its concentration. This means that current through the cell will face greater resistance in the later stages of the discharge, resulting in voltage drop.

The negative electrode, which is lead in the charged state, undergoes oxidation (the oxidation number of Pb changes from zero to plus two), forming lead ions and electrons.

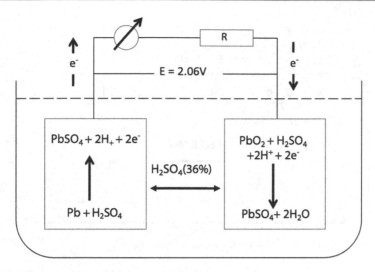

Fig. 3.1 Principle of operation of lead–acid batteries

$$Pb \leftrightarrow Pb^{2+} + 2e^- \qquad (3.1)$$

Electrons then exit the battery and pass through the external circuit, while Pb^{+2}-ions immediately react with sulfate ions from solution to form lead sulfate, $PbSO_4$, which then precipitates as crystal on the electrode. The reaction proceeds readily as there is very little concentration polarization.

$$Pb^{2+} + SO_4{}^{2-} \leftrightarrow PbSO_4 \qquad (3.2)$$

On the positive electrode, which is lead dioxide in the charged state, lead from PbO_2 undergoes reduction with protons from electrolyte and electrons from the external circuit to produce Pb^{+2}-ions and water. The oxidation number for lead changes from +4 in the dioxide to +2.

$$PbO_2 + 4H^+ + 2e^- \leftrightarrow Pb^{2+} + 2H_2O \qquad (3.3)$$

Subsequently, Pb^{+2}-ions react with $SO_4{}^{-2}$-ions to form lead sulfate, $PbSO_4$.

$$Pb^{2+} + SO_4{}^{2-} \leftrightarrow PbSO_4 \qquad (3.4)$$

The discharge reaction on both electrodes is the formation of $PbSO_4$. The overall discharge reaction involves lead and lead dioxide from the electrodes along with sulfuric acid to form lead sulfate and water. All reactions are reversed during charge. The prevailing reactions for anode, cathode, and the cell and the corresponding standard reduction potentials are summarized below:

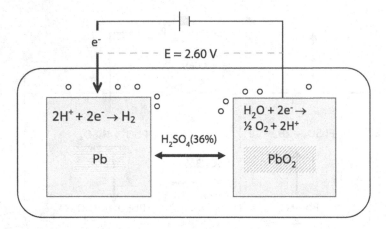

Fig. 3.2 Diagram of lead–acid battery overcharges reactions

$$Pb + HSO^-_4 \Leftrightarrow PbSO_4 + H^+ + 2e^- \quad E^\circ_{ox} = -0.355V \tag{3.5}$$

$$PbO_2 + HSO^-_4 + 3H^+ + 2e^- \Leftrightarrow PbSO_4 + 2H_2O \quad E^\circ_{red} = +1686 \text{ V} \tag{3.6}$$

$$Pb + PbO_2 + 2H_2SO_4 \leftrightarrow 2PbSO_4 + 2H_2O \quad E^\circ_{cell} = 2.041 \text{ V} \tag{3.7}$$

The equilibrium cell voltage can be calculated using the Nernst equation.

$$E = E^\circ + \frac{RT}{nF} \log \frac{[HSO^-_4]^2 \times [H^+]^2}{[H_2O]^2} = 2.041 \text{ V} + 0.059 \log \frac{[H_2SO_4]^2}{[H_2O]^2} \tag{3.8}$$

It is important to understand what happens during the charging process when a battery is already fully charged. That means all $PbSO_4$ from both electrodes is converted to lead on the negative electrode and PbO_2 on the positive electrode, but the charger or power supply is still forcing electrons from the positive electrode into the negative. Since there is no more $PbSO_4$ available, the only reactions that can take place are the hydrogen reduction or hydrogen evolution on the negative electrode and oxygen evolution on the positive electrode. Therefore, the overcharge reaction is the electrolysis of water (Fig. 3.2).

In flooded lead–acid batteries, where electrodes are immersed in liquid electrolyte, gasses generated in the overcharge reactions escape through vents at the top of battery. Prolonged overcharge causes damage, so flooded lead–acid batteries have low overcharge tolerance. Since water is consumed in the overcharge reaction, the volume and level of electrolyte decrease, exposing battery plates, which is the other term used for electrodes. This would obviously lead to loss of the active surface area and would result in lower current. In addition, vigorous gas generation from the surface of active masses can create a physical damage to electrodes by disrupting their structural consistency and causing particle detachment.

Sealed lead–acid batteries are constructed differently and have hydrogen and oxygen gases recombined inside a cell.

3.3 Types of Lead–Acid Batteries

While the majority of lead–acid batteries used to be flooded type, with plates immersed in the electrolyte, there are now several different versions of lead–acid batteries. The variations are based on several aspects, such as electrode additives, thickness of plates, variations to electrolyte, and change from open to sealed batteries.

There are two main types of batteries based on additives to electrodes: calcium and antimony batteries. The addition of 3–6% calcium makes battery plates more resistant to corrosion, overcharging, gassing, water usage, and self-discharge. All of these processes contribute to shortening the battery life. Lead–acid batteries with electrodes modified by the addition of Ca also provide for higher currents or Cold Cranking Amps. These batteries require little or no maintenance. The disadvantage is poor high-temperature performance.

The addition of antimony (chemical symbol Sb) improves the mechanical strength of electrodes, which is important for EV (electrical vehicle) and deep discharge applications, it leads to reduced internal heat and reduced water loss due to gassing. Compared to Ca addition, service life of batteries with Sb addition is greater, recharge and battery recovery from a fully discharged state are easier, and these batteries are also less expensive than the Ca version. On the other hand, addition of Sb leads to higher self-discharge (2–10% per week), compared to 1–5% per month for the calcium version, and it leads to increased gassing.

Based on the thickness of electrodes there are SLI batteries, deep discharge batteries, and forklift batteries. SLI stands for starting, lighting, and ignition; and the most common and known battery of these is the car starter battery. The car starter battery is designed to provide short bursts of high current, roughly 500 A, to start a car. This causes a battery to lose up to 5% of its charge, which is then replenished from the car alternator. The battery has thin plates or electrodes with larger surface area for high current capability. This type of lead–acid battery is designed to have high power density, but it has low total energy content and is not designed for applications that require energy delivered for long periods of time. It can also not handle deep discharge. The car battery normally operates with depth-of-discharge (DoD) of only 20%. Under those conditions, the cycle life of a car battery is around 500.

The deep discharge also called deep-cycle batteries have thicker electrodes and store more energy. The plates in these batteries are more robust and contain additives such as Ca or Sb. They are not capable of delivering high current, but the applications also don't require high current. One application is in photovoltaic systems, where a battery needs to deliver steady power for several hours during nights or periods without sun.

The golf cart or forklift batteries also contain bigger and more robust plates than the car battery, for example, they use antimony alloys to affect mechanical stability of the electrodes. They are low cost and can last for up to 20 calendar years.

Another variation of a lead–acid battery includes a different design feature—instead of battery with liquid electrolyte open to atmosphere a sealed battery with limited volume of electrolyte is made. The design prevents loss of electrolyte through evaporation, spillage, or gassing in the overcharge phase. Preventing electrolyte loss prolongs battery life. The general characteristics of sealed lead–acid batteries include improved safety because there is no free electrolyte, maintenance-free operation, and the ability to operate in any position (not possible for flooded lead–acid batteries). The electrolyte is not free, but it is gelled into moistened separators while safety valves allow venting during charge, discharge, and atmospheric pressure changes.

The small sealed lead–acid (abbreviated SLA) batteries are known as gel cells and are most commonly used in UPS or uninterruptable power supply applications. They have electrolyte in gelled form through addition of silicon dioxide, which reduces the possibility of electrolyte spillage. Gel is applied warm as a liquid and it solidifies as it cools. Cracks and voids are formed during the first few cycles, which promotes gas movement and recombination but they are not very tolerant to overcharge as they cannot handle high rates of gas evolution without risk of damage and must be charged using low current, usually C/20.

Larger batteries are called valve-regulated lead–acid (VRLA). Sealed lead–acid batteries have low overvoltage potential, which prevents gas generation during charge. Full charge is never reached in these batteries. VRLAs have pressure valves but they open only under very high pressure. Under normal operation, hydrogen and oxygen produced in the overcharge phase recombine into water on a catalyst.

One version of valve regulated lead–acid batteries uses boron silicate fiberglass mat. These are called absorbed glass mat batteries or AGMs. The glass mat is positioned between the two electrodes and it serves the purpose of preventing shorts, but it more importantly incorporates electrolyte in its pores, like a sponge. These batteries are capable of withstanding shock and vibration and no electrolyte leakage can occur even if the case is cracked or punctured. They handle gas generation extremely well and nearly all hydrogen and oxygen are recombined. Another favorable characteristic of AGM battery is that it has very low self-discharge, 1–3% per month, which makes long storage times before recharging possible. The AGM costs twice as much as the flooded version of the same capacity. Because of durability, some high-performance cars use AGMs for starter batteries instead of the flooded type.

Certain advanced lead–acid batteries are conventional, valve-regulated lead–acid (VRLA) batteries with improvements. Some of these battery systems incorporate solid electrolyte–electrode configurations such as carbon-doped cathodes, granular silica for electrolyte retention, and silica-based electrolytes. For example, carbon doping of the electrodes improves the durability and efficiency of lead–acid batteries by reducing the accumulation of lead sulfate deposits.

Other advanced battery systems incorporate capacitor technology as part of anode electrode design. They have some characteristics of capacitors such as fast response, similar to flywheels or Li–ion batteries and because of that they are sometimes called ultra-batteries or supercap batteries.

3.4 Cell Components and Fabrication

The main components of a lead–acid battery are container, active materials, grids, electrolyte, separator, and a top lid.

The battery container must be resistant to sulfuric acid and must, at the same time, provide rigid protection for electrodes and electrolyte. The material for container should be free of impurities that can affect sulfuric acid, for example, manganese and iron. The materials used for containers are hard rubber, glass, lead-lined wood, ebonite, ceramic materials, and molded plastics. The container is tightly sealed with top cover.

The construction of a container is simple, but there are a few particular features. The top cover that seals a container has holes for posts and vent plugs. In flooded batteries, a vent is used to replenish water in the electrolyte and for the escape of gasses. The bottom of the container is ribbed to hold the plates in place, especially if there is distortion that can lead to plates touching and an electrical short.

The lead–acid battery electrodes are made using two main processes: an electrochemical formation process and a "paste" process. An electrochemical process forms lead and lead dioxide through a series of charge–discharge reaction. The starting material is simply solid lead on both electrodes. The electrodes are immersed in sulfuric acid and voltage is applied. This creates water electrolysis in which oxygen is evolved on one of the electrodes, the anode. Oxygen reacts with lead and forms at first a thin layer of lead dioxide. By performing a large number of cycles, a thicker layer of lead dioxide is formed on one of the electrodes—this becomes the positive electrode. This process is called the Plante process. The negative electrode can be, in principle, made in the same way, but it is usually made through a different process called paste process. It starts with PbO, lead oxide, being pressed into the grid of a solid lead. Next, it is connected as the cathode and lead oxide are reduced to lead, forming a porous, sponge-like material with large surface area. The material is also called a paste.

The Pb grids are used as support for electrode active masses and as current collectors. The separators are made of plastic, hard rubber, fiberglass, or wood. On weight basis, lead–acid battery typically comprises 36% active materials, 27% electrolyte, 24% grids, and roughly 13% for the container, lid, and separator.

3.5 Failure Modes

One of the most important aspects of lead–acid batteries is the knowledge of possible modes of failure and how to prevent them through design and proper use.

Fig. 3.3 Schematic of battery
cross section showing loss of
electrolyte

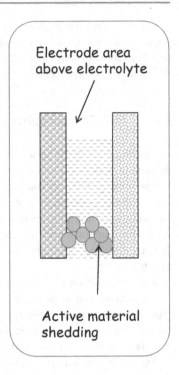

The electrolyte loss occurs in the flooded or non-sealed batteries in the overcharge phase and it is increased at high temperature and high charging rates. In sealed batteries, high charge rates cause an increase in temperature and pressure. If pressure exceeds certain preset value a valve opens and gas is released, along with some electrolyte. The loss of electrolyte means that liquid level will be lowered, exposing the top part of the plates above the electrolyte. This is shown schematically in the cell sketch in Fig. 3.3. The loss of electrolyte clearly leads to lower performance and the best way to prevent it is to properly charge the battery and avoid overcharge.

Sulfation, which means the formation of $PbSO_4$, is another serious problem with lead–acid batteries. Normally, as the lead–acid batteries discharge, lead sulfate crystals are formed on the plates. Then during charging, a reversed electrochemical reaction takes place to decompose lead sulfate back to lead on the negative electrode and lead oxide on the positive electrode. This reverse charging reaction has to take place within a certain, short period of time, about 48 h at the most. If a battery is left longer in the fully discharged state or close to fully discharged state, the lead sulfate crystals grow in size, making it impossible to breakdown such large crystals. This crystal growth occurs faster at higher temperature. It should be evident that a larger crystal, compared to more smaller crystals, means that there is a smaller surface area of the active mass in contact with electrolyte and available for reaction. The result is a reduced battery capacity. The scanning electron micrographs in Fig. 3.4 show crystal size for fully charged to close to fully discharged states.

Fig. 3.4 Scanning electron micrographs of lead–acid active mass for different stages of sulfation

Electrolyte stratification is a problem that occurs because of the specific gravity differentiation between sulfuric acid and water. In the absence of mixing or agitation, a denser sulfuric acid starts settling closer to the bottom of the battery case while water remains at the top and a concentration differential is formed. This, of course, leads to variations in electrolyte conductivity across the surface areas of the plates and the inevitable loss of performance. The gas evolution close to the end of the charging process or in the overcharge phase can have a mixing effect and help the problem of stratification, but the overcharge naturally leads to other problems. In most cases, electrolyte stratification does not lead to permanent damage because any type of agitation, including during full charge with some overcharge, mixes the electrolyte. However, during the periods of uneven electrolyte concentration a nonuniform current distribution across battery plates can occur and have an effect on the active masses. In the areas of plates toward the bottom, where the acid concentration is high, the self-discharge reaction leading to sulfate formation is faster; which then, over prolonged periods of time, may cause permanent sulfation and loss of capacity.

During extended operation at high temperature and specifically during temperature changes, the battery plates are prone to distortion. Overcharging also leads to plate distortion. Once the plates are distorted, they can puncture through the separator and cause an electrical short. This is defined as a separator failure.

All these causes of failure described so far can be largely prevented by proper battery use. Electrolyte loss can be prevented by not allowing a battery to go into the overcharge state where gasses are produced. Keeping the battery in a fully charged state and when it is discharged ensuring immediate recharge can prevent sulfation. It is obvious that these two conditions may be sometimes conflicting as there may be only a fine distinction in the chosen limitation voltage between the fully charged, not undercharged, and overcharged battery. So, it is essential to understand the battery well and, if necessary, perform detailed electrochemical characterization to determine the voltage as a function of charge profiles. Electrolyte stratification can be prevented by occasional battery agitation, while separator failure due to plate distortion can be averted if overcharge and high temperature of operation are avoided.

Besides these failures caused by operational factors, there are design failure modes that can be prevented mainly through better design and to a lesser extent through operational conditions. In flooded lead–acid batteries, roughly 85% of all failures are related to grid corrosion, while in valve-regulated lead–acid batteries, grid corrosion is the cause of failure in about 60% of cases. This is a problem that develops over time and it typically affects batteries that are close to end of life. In other words, if the preventable causes of failure are eliminated, then a battery will last close to its nominal life. The wear on the battery will finally end the battery life. The only way to extend it is to develop new materials or processes. For example, the grid in lead–acid batteries is made of solid lead and the active mass, a sponged lead for the negative electrode is pressed into the grid. The grid itself is maybe only partially exposed to electrolyte and it mainly serves as the mechanical support for the active mass and as a current collector. Over time, however, the lead in the grid slowly gets exposed to electrolyte and during the discharge process gets oxidized to form lead dioxide. There are numerous schemes to slow down grid corrosion, for example, the addition of calcium.

The extreme voltages can also cause grid corrosion. Therefore, a shallow depth of discharge reduces grid corrosion and extends battery life because the cell voltage never drops too low to values where corrosion rate is faster. Similarly, preventing overcharge, where the voltage of the positive electrode increases significantly, also minimizes chances for grid corrosion.

Another failure mode is shedding or falling of the active mass from the grid as shown in the cell sketch in Fig. 3.3. This obviously lowers the performance and since lead falls to the bottom of the container, over time it can cause a short between the two electrodes. The shedding of active mass can occur if the active material paste is not adhering well to the grid and also as a result of extreme current or temperature operational conditions.

3.6   Charge Process

The electrode potentials and cell voltage during charge and overcharge using constant current are shown in Fig. 3.5. As the charging process begins, the electrode potentials gradually increase; the positive electrode becomes more positive and negative more negative. This creates increasing potential difference and at midpoint through the charge process it is about 2.2 V. The dashed lines show the thermodynamic or theoretical electrode potentials. The difference between actual and thermodynamic potentials is called overpotential and it is higher for the positive or lead dioxide electrode.

As charging proceeds, the potentials keep gradually increasing until end of charge is reached. At this point, all lead sulfate is converted to lead on the negative electrode and to lead dioxide on the positive; and the charge is complete.

If the charging continues beyond the point of full charge, the cell potentials show pronounced, step-like increases. This is transition between the potentials required for the lead–acid reactions and the water electrolysis reactions to start. Since current from the charger or power supply is constant, it is forcing the next reactions available—which are the hydrogen and oxygen evolution. The potentials finally settle at the values required for water electrolysis.

The lower graph in Fig. 3.6 is the cell voltage or the difference between the electrode potentials and it shows the same trends described for potentials, but more

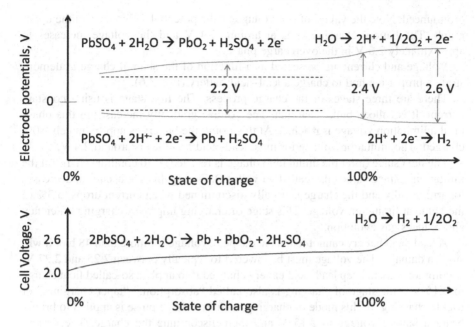

Fig. 3.5 Electrode potentials and cell voltage for a typical flooded lead–acid battery

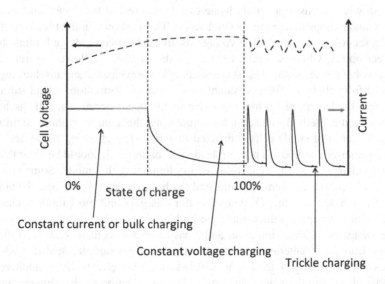

Fig. 3.6 Conceptual graph of cell voltage as a function of state of charge for a typical lead–acid battery

pronounced. Note the values of cell voltage or the potential differences in the upper graph. The battery charging ends at around 2.4 V and the voltage increases to approximately 2.6 V in the overcharge phase.

Voltage and current are presented as a function of the state of charge to demonstrate a proper method to charge a lead–acid battery (Fig. 3.6).

There are three stages of the charge process. The first stage is using constant current. It is called "bulk" charging. The voltage gradually increases in this phase until a limitation voltage is reached. At this point, the battery is approximately 80% charged. The limitation voltage for most lead–acid batteries is around 2.4 V.

The next stage (after the limitation voltage is reached) is to continue charge at the limitation voltage value (also called set voltage). During this stage, current decreases logarithmically and the charge is usually discontinued when current drops to 3% of the value at limitation voltage. This stage of charging improves charging efficiency and reduces gas evolution.

A lead–acid battery cannot remain at the peak voltage for more than 48 h or it will sustain damage. The voltage must be lowered to typically between 2.25 and 2.27 V. A common way to keep lead–acid battery charged is to apply a so-called float charge to 2.15 V. This stage of charging is also called "absorption," "taper charging," or trickle charging. In this mode of charging, a short voltage pulse is applied to briefly bring a battery voltage to 2.15 V and then discontinue the charge. The current follows by first sharply increasing and when the voltage source is removed the current logarithmically decreases. After the pulse is removed, the voltage of the battery slowly decreases; a certain hysteresis is allowed to take place and then the battery voltage drops to a predetermined value. This is shown in the trickle (or float) charging section of the graph. The voltage oscillates between voltage limitation and reconnect points, while the current similarly shows spike, followed by relaxation after the voltage is removed. This stage of charging prevents battery discharging and it keeps it fully charged. The occasional pulses prevent formation of lead sulfate or sulfation of the battery plates that otherwise would occur because of self-discharge. In addition, the method is useful for equalizing the charge without significant gassing in stacks of single cells connected in series. The charging mode described here is a preferred method of charging lead–acid batteries. It should be clear that not all battery chargers have the capability of implementing this mode. Some chargers simply apply constant current while the end of charge is determined from the voltage limitation or based on time. Of course, better chargers and most modern chargers have the ability to apply a three-stage method.

The values of voltage limitation and float or trickle voltage vary for different types of lead–acid batteries and manufacturers. For example, sealed lead–acid batteries can be charged to 2.5 V without negative effects. Any additives to electrodes also affect the voltage limitation. Proper selection of charging parameters should always be done based on the manufacturer's specifications or detailed battery evaluation using fundamental electrical characterization techniques.

The end-of-charge voltage for a single cell is close to 2.4 V and for most battery types it is more precisely 2.37 V. The most common lead–acid battery configuration on the market, the 12-V battery comprising six single cells in series, is charged with

Fig. 3.7 Capacity change versus cycle life for lead–acid battery under different charging conditions

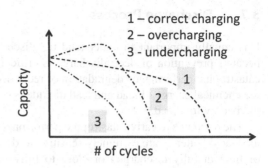

about 14.4 V and reads about 12.6 V when fully charged (in steady state, i.e., no load).

Summarizing the description from above, the proper charging occurs through a three-stage process: constant current (bulk), constant voltage (absorption), and float. As expected, the charging process has an effect on the cycle life of a battery (Fig. 3.7).

The proper charging will result in a gradual capacity decrease as a function of the number of cycles (curve 1). If a battery is constantly overcharged the capacity may for a while increase because overcharging eliminates sulfate, which is one of the important causes of capacity loss. However, after a number of cycles where the battery was forced into overcharge phase, the excessive time spent in the gassing or overcharge phase, at higher voltages, will cause premature grid corrosion and rapid drop in capacity over number of cycles (curve 2). If, on the other hand, a battery is constantly undercharged, it means that lead sulfate is never completely removed through charging. The remaining lead sulfate grows in crystal, and overtime, this becomes an unusable active mass. Furthermore, more lead sulfate is formed in each cycle since the battery is not fully charged. Over time, the capacity sharply decreases, and the battery never reaches the required or nominal number of cycles (curve 3).

It is also important to consider the effect of temperature on the charging process. The previously discussed voltage values for the voltage limitation and overcharge were all valid only for the nominal temperature of 20 °C. If temperature is different, the voltage limitation changes. At higher temperatures, the rates of reactions are higher and the limiting voltage when the charge is complete takes place sooner or at lower voltages than for nominal temperature. The gassing also starts at lower state of charge. For temperatures lower than 20 °C, the end-of-charge voltage will be higher than for the nominal temperature and gassing will start at higher state of charge.

It should be noted that all numbers and conceptual graphs represent flooded lead–acid batteries. The other types, gelled electrolyte or absorbed glass mat electrode batteries, as well as advanced lead–acid batteries will likely exhibit different behavior and voltage values will be different.

3.7   Discharge Process

It is equally important to understand the discharge reaction in lead–acid batteries because prevention of deep discharge is critical for saving the battery from early catastrophic performance degradation or reduction in battery life. During discharge, the chemical energy of lead and lead dioxide is converted to electrical by connecting the battery to a load.

The reference to early catastrophic performance degradation points to the fact that lead–acid batteries are extremely sensitive to deep discharge and prolonged periods of time in fully discharged or close to fully discharged state. It should also be remembered that these periods of time are very short—24 to 48 h. In other words, a fully discharged lead–acid battery will be irreversibly damaged if left fully discharged for more than 48 h. The cause of this is the formation of $PbSO_4$ or plate sulfation and more importantly the growth of $PbSO_4$ crystals beyond the point where they can be easily broken down to smaller crystals with regular charging methods.

There are numerous claims of techniques that can revive lead–acid plate sulfation and these techniques typically rely on some electrical or physical method of fragmenting the lead sulfate crystals. Some of the methods include high current pulses, some involve prolonged small current pulses and others may cite the use of different nonelectrical techniques. None of these techniques have been officially accepted in the battery industry and there are no systematic studies to evaluate their effectiveness. It is important to remember that conventional chargers do not have the capability to reverse the plate sulfation and recover the battery performance.

A typical voltage curve is given for a sequence of discharge, followed by a charge process (Fig. 3.8).

The discharge portion of the curve indicates that voltage does not stay flat for most of the discharge. Rather, it exhibits gradual voltage decrease and a rapid drop at the end of discharge. The recommended end of discharge voltage is 1.75 V/cell, which should be at the knee of the curve, before rapid decline. The fact that discharge

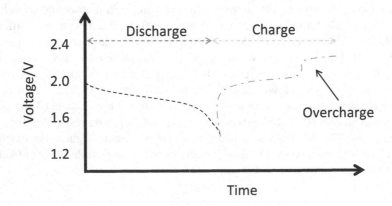

Fig. 3.8 Voltage versus time for typical lead–acid battery discharge and charge

Fig. 3.9 Cell voltage versus depth of discharge for different discharge currents

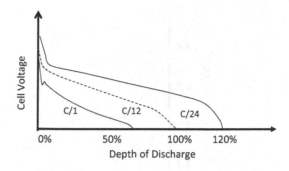

does not follow a preferred flat curve has to be considered as it may have an effect on the load or a device powered by a battery. It is essentially important to make sure that voltage at any point during discharge will be sufficient for a given application.

During the discharge process, the electrodes become coated with lead sulfate and acid electrolyte becomes weaker. The shape of the voltage discharge curve depends on the discharge current (Fig. 3.9).

Voltage decreases steeper for higher discharge current and the nominal battery capacity in ampere-hours (Ah) can be expected only for nominal discharge current, which is typically C/14 to C/12. For currents higher than nominal, less than 100% of the stated nominal capacity will be obtained; and for currents lower than nominal, capacity in excess of nominal is expected.

3.8 Electrolyte

It has already been established that electrolyte participates in the electrochemical reactions in a lead–acid battery. In the discharge reaction, the acid is consumed, and it participates in forming lead sulfate. In the process, the acid concentration or its specific gravity is reduced (Fig. 3.10). The graph shows the specific gravity of sulfuric acid during charge and discharge. It first gradually decreases during discharge, then it rises during charge.

As a battery is charged, the lead sulfate coating on the electrodes is removed and acid electrolyte becomes stronger. In the overcharge phase, shown as the last portion of the graph, a concentration or the specific gravity, higher than the initial concentration can occur. The reason for this is that hydrogen and oxygen are generated in the overcharge reaction, consuming water from the electrolyte and effectively increasing the acid-specific gravity. In flooded lead–acid batteries, water is replenished through a cap and concentration is lowered back again.

Since the concentration of sulfuric acid changes based on the state-of-charge and voltage changes with the state of charge, it is expected that there is a direct correlation between the cell voltage and sulfuric acid concentration (Fig. 3.11).

At a certain point in the life of a battery, acid concentration becomes too low, which results in lower voltage and lower current due to reduction in conductivity. The battery capacity is also diminished at that time. This is the point when it is

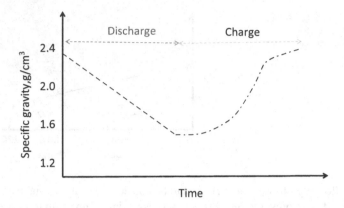

Fig. 3.10 Sulfuric acid–specific gravity changes during discharge and charge processes

Fig. 3.11 Cell voltage versus electrolyte-specific gravity

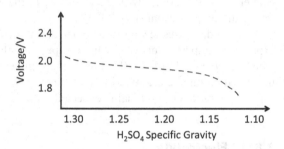

recommended that battery use be discontinued and the battery recycled. The electrolyte conductivity of sulfuric acid changes with its density (Fig. 3.12).

A maximum conductivity is observed for the acid-specific gravity of roughly 1.3 g/cm^3. Since the conductivity varies with concentration, this means that the internal resistance of lead–acid batteries changes during charge and discharge processes and is the lowest around the midpoint in the discharge or charge. In the later stages of discharge, when the battery is approaching full discharge, the conductivity decreases and limits the current that the battery can deliver.

A further complicating factor is that the specific gravity of sulfuric acid depends on temperature (Fig. 3.13).

The specific gravity at 15 °C is used as the reference value and specific gravity at other temperatures is calculated using a temperature coefficient. The change in specific gravity with temperature has two main implications. First, the conductivity of electrolyte also changes with temperature and obviously the performance of a battery is reduced at lower temperatures. Second, because the electrolyte is water based, the performance of a battery is critically affected at very low temperatures and in extreme cases, the electrolyte can freeze. Obviously, if electrolyte freezes, the conductivity is essentially reduced to zero and the battery ceases its performance. Moreover, electrolyte freezing around the plates causes damage and even upon

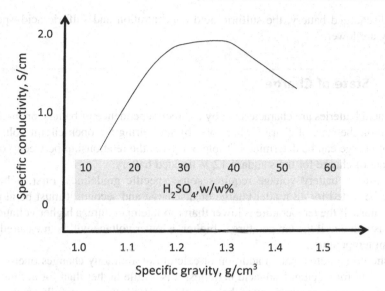

Fig. 3.12 Conductivity of H_2SO_4 versus acid concentration

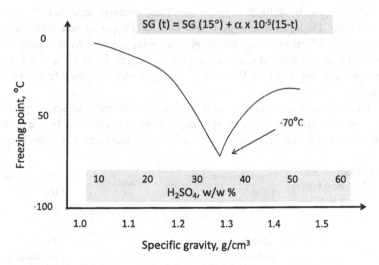

Fig. 3.13 Freezing point of H_2SO_4 versus acid-specific gravity

warming up and melting the electrolyte, permanent damage is likely. The freezing temperature depends on the concentration, but there are some complex effects that make the dependence nonlinear, as seen in the graph above. It can also be seen from the graph in Fig. 3.13 that the lowest freezing temperature is for acid-specific gravity of around 1.3 g/cm³. A comparison of concentration scales shows that this would be equivalent to a molality of 5–6 and acid weight percent of 30–40. This concentration of sulfuric acid is characteristic of a nearly fully charged battery. For partially or fully discharged battery, the sulfuric acid concentration and sulfuric acid–specific gravity are lower.

3.9   State of Charge

Lead–acid batteries are characterized by a direct dependence of battery open-circuit voltage on the state of charge. Therefore, by measuring the open-circuit voltage, a state of charge can be determined. Table 3.1 gives the relationship between voltage and state of charge for the standard 12 V flooded battery.

Measuring battery voltage requires some specific guidelines. First, a battery should be rested for 48 h after charge or discharge and second, it must be at room temperature. If the temperature is lower than room temperature, a higher voltage will be measured and if the temperature is higher, a lower voltage will be measured than at room temperature.

It should be noted that an addition of calcium or antimony changes open-circuit voltage and for calcium lead–acid batteries, it is a little higher than for antimony.

Table 3.2 gives an overview of the several battery parameters in fully charged and fully discharged states.

State of charge is by definition 100% in a fully charged state and 0% in a fully discharged state. Depth of discharge is 0% in a fully charged and 100% in a fully discharged state. Electrolyte concentration is approximately 6 M in a fully charged and 2 M in a fully discharged state, while its specific gravity is 1.3 and 1.1. Finally, the open-circuit voltage is 12.7 V in a fully charged and 11.9 for a fully discharged 12 V battery.

The internal resistance changes with the state of charge and since open-circuit voltage is an indication of the state of charge, there is correlation between internal resistance and open-circuit voltage. The resistance increases with discharge and

Table 3.1 Open circuit voltage of lead–acid battery versus state of charge

Open-circuit voltage	State of charge, %
12.65 V	100
12.45 V	75
12.24 V	50
12.06 V	25
11.89 V	Discharged

Table 3.2 Battery parameters for fully charged and fully discharge states

Parameter	Fully charged	Completely discharged
State of charge	100%	0%
Depth of discharge	0%	100%
Electrolyte concentration	~6 M	~2 M
Electrolyte-specific gravity	~1.3	~1.1
Open-circuit voltage	12.7 V	11.9 V

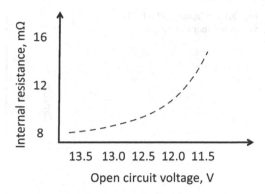

Fig. 3.14 Internal resistance of LAB versus open-circuit voltage

decreases with charge, it also follows changes in electrolyte concentration (the more dilute acid in the discharge state means lower conductivity). The resistance increase is almost linear with a decrease in specific gravity and a decrease in open-circuit voltage from 13.5 to 11.5 V Fig. 3.14).

Resistance roughly doubles from fully charged to fully discharged state. Temperature affects internal resistance (shown in the graph in Fig. 3.14 at room temperature). Cold temperatures increase internal resistance for all batteries, including lead–acid batteries.

3.10 Capacity

One of the most important properties of lead–acid batteries is the capacity or the amount of energy stored in a battery (Ah). This is an important property for batteries used in stationary applications, for example, in photovoltaic systems as well as for automotive applications as the main power supply. Capacity is less critical for car starter batteries where the power or current delivered is more important. The requirements of an application are reflected in the construction of lead–acid batteries relative to the thickness of its plates. For power or current batteries, such as automotive starter batteries, plates are thinner and with larger surface area; while for energy applications, plates are thicker, with more active mass, but less surface area readily accessible for the reaction.

The plates for the automotive starter batteries are about 0.040″ (1 mm) thick, while typical golf cart batteries have plates that are between 0.07 and 0.11″ (1.8–2.8 mm) thick. Forklift batteries may have plates that are much thicker and exceed 0.250″ (6 mm).

The capacity critically depends on the temperature. Because chemical reactions in batteries proceed faster at higher temperatures and internal resistance of a battery to current flow is lower, the capacity is also higher at higher temperature and lower at lower temperature. The capacity is 100% for a new battery at nominal temperature and it is obvious that the capacity can be higher or lower than the nominal value depending on the temperature.

Fig. 3.15 Capacity of a LAB
versus number of cycles

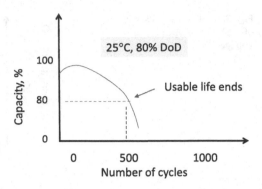

The capacity considered here is instantaneous, but there is also a secondary effect of temperature on battery life. At lower temperatures, the battery life is extended and at higher temperatures it is shortened. This has implications on the design of a battery for specific applications, such as photovoltaic systems. If the application is at temperatures other than room temperature, 25 °C, battery sizing must be done to compensate for capacity loss or gain. As the temperature decreases from room temperature to 50 °F (10 °C) and then to 20 °F (−6.7 °C), the multiplication factors of 1.19 and then 1.59 must be used to resize the battery.

The capacity generally declines with the number of cycles and over the life of a battery (Fig. 3.15).

There is usually a slight increase within the first 100 cycles due to active mass redistribution, but then a steady decline of capacity with cycle life occurs. The slope depends on the depth of discharge as well. The end of life is usually considered when the battery capacity drops to 80% of the initial value. For most lead–acid batteries, the capacity drops to 80% between 300 and 500 cycles.

3.11 Cycle Life

Lead–acid battery cycle life is a complex function of battery depth of discharge, temperature, average state of charge, cycle frequency, charging methods, and time. The rate of self-discharge also plays a role. In general, as for all other batteries, the cycle life decreases with an increase in depth of discharge and temperature (Fig. 3.16).

Cycle life increases at lower temperature of operation and storage and with shallow discharge cycles. For lead–acid batteries, the depth of discharge should be less than 80%, if cycle life is important.

The depth of discharge is the critical operational condition affecting cycle life. The deeper the depth of discharge, the more $PbSO_4$ is formed and it may not always be broken down to smaller crystals during charging. Over time and number of cycles, more and more residual lead sulfate remains in the active mass. Other degradation mechanisms also take place and they are enhanced at deep depth of

Fig. 3.16 LAB cycle life versus depth of discharge for different temperatures

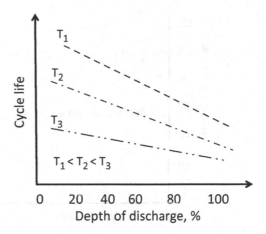

discharge. These mechanisms are grid corrosion of the positive plate, depletion of active material, and expansion of the positive plate.

For lead–acid batteries, a typical number of discharge/charge cycles at 25 °C (77 °F) with respect to the depth of discharge is:

- 150–200 cycles with 100% depth of discharge (full discharge)
- 400–500 cycles with 50% depth of discharge (partial discharge)
- 1000 and more cycles with 30% depth of discharge (shallow discharge)

It is generally considered and used in design for most applications that the end of life of a battery is when 20% of capacity is lost.

It is also clear that elevated temperature reduces longevity. For flooded lead–acid batteries and for most deep-cycle batteries, every 8 °C (about 15 °F) rise in temperature reduces battery life in half. For example, a battery that would last for 10 years at 25 °C (77 °F) will only be good for 5 years at 33 °C (91 °F). Theoretically, the same battery would last a little more than 1 year at a desert temperature of 42 °C.

The service life of a lead–acid battery can in part be measured by the thickness of its positive plates. During charging and discharging, the lead on the plates gets gradually consumed and the sediment falls to the bottom. As a result, the measurement of the plate thickness can be an indication of how much battery life is left. The weight of a battery is also a good indication of lead content and life expectancy.

The available capacity is impacted by the depth of discharge and is also a function of the number of cycles. After each cycle, a small portion of the battery active mass becomes sulfated or grid corrosion occurs. These are irreversible processes that lead to a gradual loss of capacity in Ah compared to the first cycle. It is wrong to assume that a certain depth of discharge would mean a delivery of the same capacity over a number of cycles. In the later stages of the battery life, less capacity will be delivered, even though a depth of discharge may indicate the same value. Since

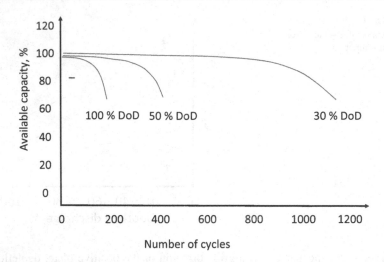

Fig. 3.17 Available capacity of a LAB versus number of cycles for different depth-of-discharge values. (Adapted from various references)

capacity is measured versus the previous cycle it would mean that it slowly decreases compared to the first cycle.

In Fig. 3.17, the percentage of available capacity versus number of cycles a battery can reach in its lifetime is given for a practical system. The end of life was chosen as 60% of the initial capacity for all three cases.

The first curve from the left shows what happens if a lead–acid battery is discharged fully each cycle or the depth of discharge is 100%. The maximum cycle life that a battery can reach before the capacity drops to 60% is around 200. The curve for 50% depth of discharge shows a similar trend with the maximum number of cycles between 500 and 600. This more gradual decrease results in higher number of cycles.

Finally, at 30% depth of discharge, a lead–acid battery experiences fairly constant capacity, around 100% of the initial for most of the lifetime. Because this is very shallow discharge mode, a battery lasts much longer than the nominal capacity and can reach over 1000 cycles. When it finally reaches its end of life, the available capacity drops to 60%.

The data in Fig. 3.18 are an attempt to combine three effects that, while they have no three-way direct correlation, all contribute to affecting battery life. In the graph, a battery life in years (not cycle life) is shown versus depth of discharge for three batteries with different number of cycles per year. First, the battery life in years declines with the depth of discharge, which is expected. And, the more cycles per year that battery goes through the shorter its lifetime in years. It can also be seen from the graph that the lines for different number of cycles per year become slightly steeper.

Besides the obvious that can be obtained through simple calculation of cycles per year multiplied by the number of years, there are several mechanisms at play. We

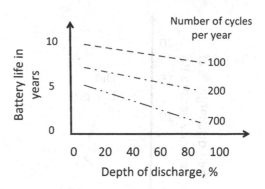

Fig. 3.18 LAB life in years versus depth of discharge for different number of cycles per year

know already that battery is limited by number of cycles and it reaches its lifetime when some roughly nominal number of cycles is reached, depending how many cycles the battery has per year. But the side chemical reactions proceed often regardless of the electrical cycling. They usually result in material degradation such as corrosion and contribute to shortening the life in years or the calendar life of a battery. The effects of electrical charging and discharging are combined with material degradation effects to give a true measure of how long a battery can last.

3.12 Self-Discharge

Another important performance factor for lead–acid batteries is self-discharge, a gradual reduction in the state of charge of a battery during storage or standby. The self-discharge takes place because of the tendency of battery reactions to proceed toward the discharged state, in the direction of exothermic change or toward the equilibrium. The discharge state is more stable for lead–acid batteries because lead, on the negative electrode, and lead dioxide on the positive are unstable in sulfuric acid. Therefore, the chemical (not electrochemical) decomposition of lead and lead dioxide in sulfuric acid will proceed even without a load between the electrodes.

Self-discharge reactions in LAB proceed faster at higher temperatures, so storing them at lower temperature reduces the loss of capacity through self-discharge. It is commonly accepted that most lead–acid batteries have about a 5% self-discharge rate, which means they lose 5% of their capacity per month, at 20 °C (Fig. 3.19).

As with other operational factors for lead–acid batteries, self-discharge is also a result of complex interactions and the rate of self-discharge depends on battery configuration, additives, but also on the history of a battery prior to storage relative to depth-of-discharge and temperature. The age of a battery is also a significant factor contributing to self-discharge and the rate increases with the age and cycle life.

Additives play a crucial role in reducing the rate of self-discharge (Fig. 3.20).

The rate of self-discharge is significantly lowered by the addition of Sb to Pb and even further lowered by the addition of Ca. As with all other battery-operating

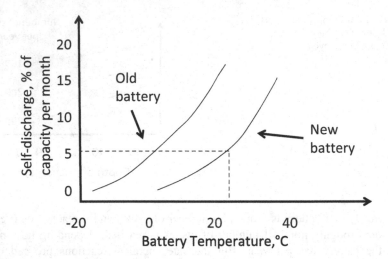

Fig. 3.19 Lead–acid battery self-discharge as a function of temperature for new and old batteries

Fig. 3.20 Capacity of LAB versus days of storage for electrodes with additives

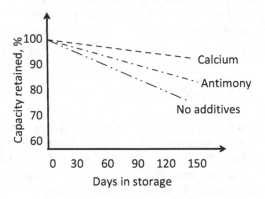

Fig. 3.21 Daily reduction in the acid-specific gravity as a function of temperature

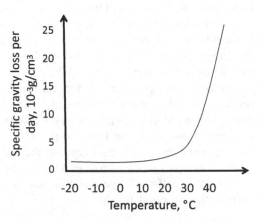

characteristics, the data in the graphs are not specific for a certain battery and is representative or average. Specific batteries may have different values.

The self-discharge reactions produce $PbSO_4$ and sulfuric acid is consumed in this reaction. As a result, the concentration of sulfuric acid and its specific gravity decline during storage due to self-discharge reactions (Fig. 3.21). The specific gravity loss increases at higher temperatures and this is naturally a result of a faster self-discharge reaction.

Nickel–Cadmium Batteries

4

4.1   Overview and Characteristics

Batteries using nickel negative electrodes are commonly called nickel-based batteries or simply nickel batteries. The first commercial battery system based on nickel electrode was nickel–cadmium, invented in 1899. The nickel–cadmium battery is an exceptional battery, but often neglected when selecting a battery for an application because of the lack of understanding. For poorly informed system designers, the knowledge of batteries is limited and they often easily decide on a standard choice such as lead–acid battery or a newly very popular lithium–ion battery. However, nickel–cadmium batteries are very attractive for many applications and their performance makes them superior for many conditions.

Batteries with nickel oxyhydroxide positive electrode are very popular batteries with alkaline electrolyte. The nickel electrode, which has layered structure, can be paired with cadmium, iron, zinc, metal hydride, and even hydrogen negative electrodes.

Nickel–cadmium battery was invented in 1899 by Waldemar Jungner from Sweden. The first sealed version was accomplished in 1947 by Neumann and this paved the way to modern nickel–cadmium batteries.

The advantages of nickel–cadmium batteries are high number of cycles (typically over 1000), better energy density than lead–acid batteries, low internal resistance and high power density, good performance at low temperatures, long shelf life, and fast recharge.

However, nickel–cadmium batteries have low energy density compared to nickel–metal hydride and lithium–ion batteries. Another apparent disadvantage of nickel–cadmium battery is the so-called memory effect which makes periodical full discharge necessary. Because of cadmium toxicity, nickel–cadmium batteries are considered environmentally unfriendly and problematic. For this reason, nickel–cadmium batteries are as of lately restricted in the European Union countries. Other

© The Editor(s) (if applicable) and The Author(s), under exclusive license to   73
Springer Nature Switzerland AG 2021
S. Petrovic, *Battery Technology Crash Course*,
https://doi.org/10.1007/978-3-030-57269-3_4

disadvantages of nickel–cadmium battery are the high rate of self-discharge, poor performance at high temperatures, and complex charging.

4.2   Principle of Operation

During discharge reaction in N–Cd battery, cadmium is oxidized on the negative electrode to form $Cd(OH)_2$ and electrons (Fig. 4.1).

On the positive electrode, nickel oxyhydroxide (NiOOH) decomposes to form nickel hydroxide ($Ni(OH)_2$) and hydroxyl ions (OH^-), which replenish OH^- consumed in the oxidation reaction. As a result, the electrolyte, which is 21% potassium hydroxide, is not changed in the reaction, like sulfuric acid in lead–acid batteries, because there is effectively no hydroxide being consumed in the reaction.

The reactions are reversed on charge and until active masses in discharged states $Cd(OH)_2$ and $Ni(OH)_2$ are available, the reactions proceed as expected. At the end of charge, the battery is rarely immediately disconnected from charger or power supply and the battery reactions enter an overcharge phase. The overcharge is an undesirable process in Ni–Cd batteries because it leads to generation of gasses and increase in both pressure and temperature that can catastrophically damage a battery. Since most nickel–cadmium batteries are sealed, a special design approach was needed to control the overcharge and to prevent any damage to battery. The solution was found through the use of oversized negative electrode (Fig. 4.2).

At the end of the charging process, oxygen is evolved on the positive electrode and hydrogen on the negative. If negative cadmium electrode is oversized in capacity, then nickel electrode (the positive electrode) reaches full charge before cadmium electrode. As a result, oxygen is generated before hydrogen generation

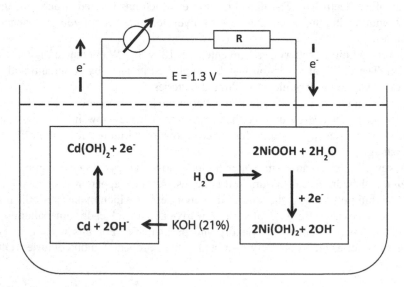

Fig. 4.1 Schematics of N–Cd cell and reactions during charge process

Fig. 4.2 Schematics of Ni–Cd battery and reactions during the overcharge phase. The part of the active mass capacity that is in excess is shown in gray

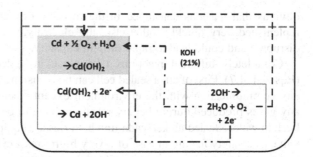

starts. Evolved oxygen diffuses from nickel electrode to cadmium electrode and reacts there in a chemical reaction with elemental cadmium to form cadmium hydroxide ($Cd(OH)_2$). This is shown with dashed connecting line (----) in the sketch. Since cadmium is produced in an electrochemical reaction and consumed in a chemical reaction with oxygen, a cycle is established and cadmium electrode can never reach full charge where it would produce hydrogen. Because oxygen is consumed in recombination reaction, pressure never rises to create a problem. The recombination does not always proceed at fast enough rate and additional design features have to be used on the cadmium active mass.

4.3 Reactions During Overcharge

Chemical equations for the reactions that occur in open or vented cells and in sealed cells are shown below.

Vented (open cells)

$$\text{Positive electrode}: \quad 4OH^- \;\rightarrow\; 2H_2O + O_2 + 4e^- \tag{4.1}$$

$$\text{Negative electrode}: \quad 4H_2O + 4e^- \;\rightarrow\; 2H_2 + 4OH^- \tag{4.2}$$

$$\text{Net reaction}: \quad 2H_2O \;\rightarrow\; 2H_2 + O_2 \tag{4.3}$$

Sealed cells

$$\text{Positive electrode}: \quad 4OH^- \;\rightarrow\; 2H_2O + O_2 + 4e^- \tag{4.4}$$

$$\text{Negative electrode}: \quad 2Cd(OH)_2 + 4e^- \;\rightarrow\; 2Cd + 4OH^- \tag{4.5}$$

$$\text{Net(electrochemical)} \quad 2Cd(OH)_2 \;\rightarrow\; 2Cd + 2H_2O + O_2 \tag{4.6}$$

$$\text{Chemical recombination} \quad 2Cd + O_2 + 2H_2O \;\rightarrow\; 2Cd(OH)_2 \tag{4.7}$$

In vented cells, the overcharge phase reactions, after all active masses have been converted to their forms in fully charged states, switch to water electrolysis reactions. On the positive nickel electrode, oxygen is produced and on the negative, cadmium electrode hydrogen is produced. The two gases escape through a vent. It is obvious that water is consumed in the reaction of electrolysis, so it needs to be replenished very quickly; otherwise, the electrolyte and active mass balance are disrupted and could lead to rapid drop in capacity.

Completely different reactions take place in sealed cells during overcharge (Eqs. 4.4–4.7). First of all, a sealed cell can be envisioned as a closed system without any communication with the environment except electrical signal and heat, hence any gases produced during battery reactions have to be contained and recombined back to water. Sealed nickel–cadmium batteries are equipped with a pressure release valve as a safety feature in case of a very high-pressure buildup.

Gas evolution takes place primarily in the overcharge phase, but to some extent as well in case of rapid charging or discharging. As explained earlier, the negative cadmium electrode is oversized, so the positive nickel electrode reaches full charge before the cadmium electrode. This means that all nickel hydroxide has been converted to nickel oxyhydroxide (NiOOH). After that, the only oxidation reaction that can take place is oxygen evolution from hydroxyl ions (Eq. 4.4). Note that this is an electrochemical reaction, producing electrons, which are drawn by the power supply to the negative electrode.

At the negative electrode, $Cd(OH)_2$ is reduced to elemental cadmium in a standard charge reaction, with the generation of OH^-. Note again, that there is no net consumption of OH^- ions since they are produced on the negative electrode and consumed on the positive. Equation (4.6) shows the overall or sum reaction of two electrochemical, redox reactions. Electrons are not explicitly shown in the overall reaction because the same number of electrons is involved in each redox reaction.

Finally, oxygen produced on the positive electrode first fills the space in the cell above the electrodes and in between the electrodes, then diffuses through the electrolyte and separator and reaches the negative cadmium electrode. There, oxygen reacts with elemental cadmium (that has been just formed in the electrochemical reduction reaction) and water. The result of this chemical, not electrochemical, reaction is that elemental cadmium is converted back to cadmium hydroxide (Eq. 4.7).

In summary, cadmium hydroxide is consumed in the electrochemical reaction and produced in the chemical reaction. If rates of these two reactions are properly matched, the amount of cadmium hydroxide in the negative electrode is balanced and stays approximately the same as at the start of oxygen evolution on nickel electrode. In this way, the negative cadmium electrode never reaches full charge and hydrogen is never produced. The goal is naturally to prevent hydrogen from forming because in sealed environment the presence of hydrogen and oxygen under pressure could easily lead to a violent reaction and explosion.

4.4   Voltage During Charge and Discharge

General voltage trend during charge and discharge for nickel–cadmium battery is shown in Fig. 4.3.

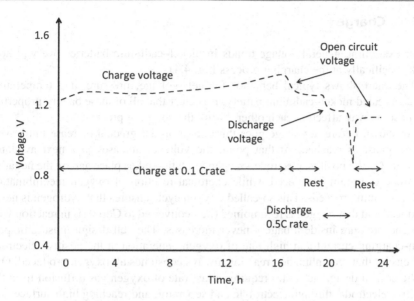

Fig. 4.3 General trend for cell voltage during charge and discharge in a Ni–Cd battery

Voltage increases during charge from 1.2 V, which is the nominal cell voltage for Ni–Cd cells. It is worthwhile noting here that this voltage is lower than for alkaline zinc–carbon primary cells and therefore nickel–cadmium is not an adequate replacement for just any application. However, many electronic devices require around 1.0 V. The alkaline cells have the initial or full-charge voltage of 1.5 V and then during discharge the voltage steeply drops to an average voltage of 1.2 V. The nickel–cadmium battery voltage is very stable, it changes very little during discharge and remains around 1.2 V. It is clearly beneficial for voltage to be stable during discharge; however, this makes it impossible to detect battery state-of-charge by measuring its voltage.

Cell voltage rises gradually during charge as shown in the graph in Fig. 4.3 for C/10 rate. At the end of charge there is a steeper increase beyond 10 h, as shown in the figure. This is the overcharge phase where cell voltage follows a steeper increase and plateau. This voltage corresponds to new reactions taking place and, in sealed nickel–cadmium batteries, these are oxygen evolution on the nickel electrode and cadmium reduction on the negative electrode. When the charge is discontinued, the cell voltage relaxes back to approximately 1.2 V.

A discharge, shown in Fig. 4.3, starts at C/2 rate and in 2 h the voltage drops to close to 1.0 V which means the end of discharge. If the discharge continues beyond cutoff voltage, as shown in the chart, the voltage sharply drops because the capacity is nearly fully exhausted. This condition has to be avoided in the application, but it may occur with a heavy load or at high discharge current. If the load is now removed, the cell voltage recovers and settles between 1.1 and 1.2 V.

4.5   Charge

After examining general voltage trends in nickel–cadmium batteries, we will now look specifically at the charging process Fig. 4.4).

The chart shows typical behavior for cell voltage, pressure, and temperature inside a sealed nickel–cadmium battery. It is clear that all of these battery properties are in some way affecting each other during the charging process.

As shown before, a voltage goes through an initial gradual increase until over-charge phase is reached. At that point, the voltage increases to a next available reaction. On the positive electrode, oxygen evolution takes place and on the negative electrode cadmium is reduced, while chemical reaction of oxygen recombination with cadmium proceeds. This so-called oxygen cycle ensures that hydrogen is never produced and that oxygen is recombined (i.e., converted to $Cd(OH)_2$ in reaction with Cd) and pressure inside a battery never increases. The cell design must anticipate high-charging current and high rate of oxygen generation at the positive electrode and ensure that recombination reaction always consumes all oxygen produced. One of the critical design attributes requires a fast rate of oxygen gas diffusion from the positive electrode, through electrolyte and separator, and reaching high surface area of the negative active mass in contact with electrolyte. In such a configuration, the reaction takes place on a so-called three-phase boundary between the solid active mass, liquid electrolyte, and gaseous oxygen. If the rate of oxygen evolution is faster than that of recombination, gas starts accumulating in a cell and increasing pressure must be released through a valve.

As the charging process in a sealed Ni–Cd battery advances, temperature rises because of heat released in the charging reaction. The temperature rise is very critical and must remain low during charging to prevent performance degradation.

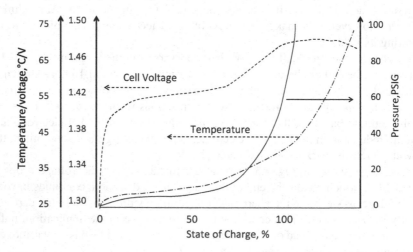

Fig. 4.4 Generalized cell voltage, temperature, and pressure for hypothetical Ni–Cd battery during charge as a function of state of charge

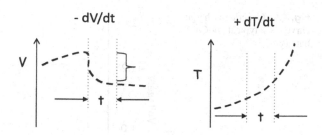

Fig. 4.5 Charge indicators for Ni–Cd batteries, voltage change over time on the left and temperature change with time on the right

The charge efficiency depends on the charging current and is somewhat surprisingly higher for faster charging currents (90% for 1 C current and 70% for 0.1 C current). Fast charging is therefore beneficial for nickel–cadmium batteries. Slow charging using low currents allows crystal growth that leads to problems generally described as memory problems (see below). Slow charging, typically for 24 h, is used only for the very first time the battery is used in order to equalize the cells in a pack; and after prolonged storage, a trickle charge with low current must be used to redistribute the electrolyte, which tends to segregate on the bottom of the cell.

Neither the voltage of a battery nor the temperature shows distinct characteristics to help determine the end of charge. By default, the charge is time-controlled and typically a battery would be left on a charger longer, creating potentially dangerous conditions of extreme temperature or pressure. More sophisticated charging systems have the ability to detect the full charge through combination of voltage drop at full charge (negative delta dV/dt), rate of temperature increase (dT/dt), absolute temperature, and timeout timers (Fig. 4.5).

Voltage relaxation and temperature increase are measured for a certain period of time in seconds to determine the end-of-charge point. Advanced chargers use high current at the beginning and then lower current after 70% of charge to achieve better charge acceptance.

Fast charging is a preferred method for charging Ni–Cd batteries, but it should be applied with good monitoring and control of voltage, temperature, and pressure to prevent overcharging and the creation of potentially hazardous conditions; 1 C charging rates are common for nickel–cadmium batteries and 4–6 C charging rates are also often used, charging a battery in 10–15 min. At a high charging rate such as 4 C, the amount of heat generated is 16 times higher than at 1 C rate because the internal resistance and therefore the heat generated are proportional to the square of the current.

4.6 Discharge

Nickel–cadmium batteries, unlike some other battery systems, show very stable voltage of 1.2 V for the majority of the discharge process up to the point where there is a "knee" in the curve and a sharp drop at the end of discharge (Fig. 4.6). The point when the battery reaches 0.9 V is considered the end of discharge and full capacity.

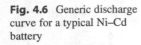

Fig. 4.6 Generic discharge curve for a typical Ni–Cd battery

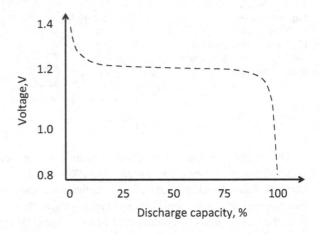

Fig. 4.7 Battery capacity as a function of discharge current expressed as C-rate. The values are hypothetical and typical

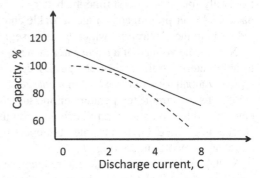

Stable discharge voltage in Ni–Cd batteries is attainable for most discharge currents, except for very high C-rates, where a sloped curve can be observed. It is hard to quantify that C value because it depends on the particular battery type. In general, the shape of a curve and discharge characteristics depend on the discharge current, depth of discharge, cell temperature, rest time after charge, and previous battery history.

The capacity that can be obtained from a battery depends on the discharge current (Fig. 4.7).

Most manufacturers quote nominal or 100% capacity at 1 C currents, which is considerably higher nominal current than for other battery systems such as lead–acid. The dependence is not always linear (solid line) and it can also have a shape similar to the dashed-line curve, where at lower currents, typically less than C/2, the capacity varies little with the C-rate. It should be noted that the capacity can be higher than nominal, for a discharge current less than nominal.

4.7   Effect of Temperature on Discharge

The effect of temperature on the discharge performance of Ni–Cd batteries is similar to the typical behavior of batteries in general. At higher temperatures, the reaction rates are faster and larger capacity can be obtained, while at lower temperatures, the processes are slower, internal resistance is higher, and lower capacity is obtained (Fig. 4.8).

It can be seen in these conceptual (not real data) graphs that the shapes of voltage curves are still very flat for the majority of discharge, with somewhat steeper slope close to the knee when the battery is approaching zero state of charge and voltage is nearing the cutoff voltage of 0.9 V. The curves for different temperatures show lower capacity at 0 °C and −20 °C than at 20 °C, consistent with the general behavior of batteries.

A more complex situation develops at higher discharge currents (e.g., 8 C) and temperatures higher than nominal of 25 °C (Fig. 4.9). Two factors affecting battery voltage (and capacity) are discharge current and temperature, which cause opposite effects—discharge current increase lowers the voltage (and capacity) while higher temperature increases both. A conceptual example in the figure predicts the

Fig. 4.8 Conceptual trend in battery voltage during discharge at 0.2 C rate as a function of discharge capacity for different temperatures

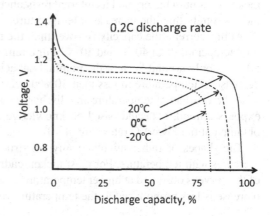

Fig. 4.9 Battery voltage during discharge at 8 C rate as a function of discharge capacity for different temperatures. Discharge curves conceptually constructed based on assorted real battery data

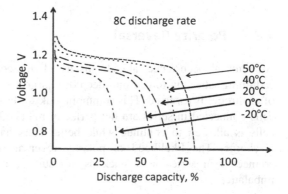

Fig. 4.10 Battery capacity as a function of temperature for different discharge currents. Discharge curves conceptually constructed based on assorted real battery data

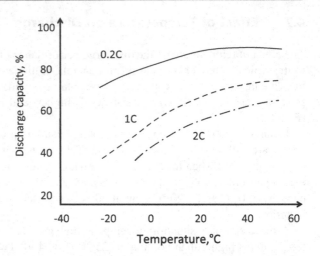

prevailing effect of high discharge current, resulting in capacity less than 100% of the nominal even for temperature of 50 °C, which alone should have led to increase in capacity. However, a rather high discharge current of 8 C tests the limits of the nickel–cadmium battery and its internal resistance; it affects the voltage and capacity more strongly than the increase in temperature.

At this current, the capacity is lower than the nominal capacity of 100% even for high temperatures of 40 °C and 50 °C. For example, capacity that can be obtained at 50 °C is only around 80% of the nominal capacity and it keeps falling with further decrease in temperature to less than 40% at −20 °C.

A direct effect of temperature and discharge current (or C rate) on battery capacity exposes a trend, expected based on knowledge so far, that the highest capacity is obtained at the lowest C rates (Fig. 4.10).

The shapes of rather hypothetically constructed curves show that the capacity increases with temperature. For most nickel–cadmium batteries, a gradual flattening of the curve is observed at higher temperatures, which means that the rate of capacity increase is less pronounced as the temperature rises above the nominal 20 °C.

4.8   Polarity Reversal

An unusual occurrence, already discussed in Sect. 4.2, transpires when nickel–cadmium battery is forced into deep discharge (or overdischarge) and the polarity of electrodes is reversed. This commonly takes place in battery stacks (i.e., packs) if cells connected in series are not perfectly matched and discharge through weaker cells is allowed to continue while better cells have not yet reached the end of discharge. The likelihood of polarity reversal increases when more cells are connected in series in a stack as a result of capacity and internal resistance imbalance.

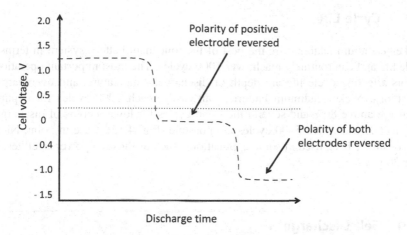

Fig. 4.11 Battery voltage of a single cell versus discharge time during polarity reversal. The voltage values are approximate and assumed based on variety of actual battery data

As voltage of a single cell in a stack is pushed below the cutoff voltage of 0.9 V, it rapidly drops to below zero because the positive electrode receives more negative charge, i.e., electrons that have nowhere to go, but can collect on the electrode and reverse its polarity (Fig. 4.11). The positive electrode is usually the first to run out of capacity.

An additional complication in this case is the evolution of hydrogen at what used to be the positive electrode and since there is no mechanism for hydrogen recombination, a pressure buildup is inevitable. As the discharge continues, the potential reversal of the previously negative electrode ensues as well, and voltage is further reduced to about -1.4 V. If the cell exhibiting polarity reversal is further discharged with current forced by other cells to flow in the original direction, it produces an electrical short, which is deleterious for the whole stack.

A solution described in 4.2 for preventing the problem is found in the form of so-called antipolar mass. The antipolar mass is a small amount of $Cd(OH)_2$ added to the positive electrode—in order to affect the processes during "overdischarge." Since reduction occurs in the discharge on the originally positive electrode, the anti-polar $Cd(OH)_2$ mass converts to Cd. At the same time, oxygen is now produced on the originally negative electrode (cadmium electrode) in the oxidation reaction and it reacts with Cd in the originally positive electrode to convert back to $Cd(OH)_2$. In this way, the problem of gas evolution in the overdischarge phase is being controlled, but it should be understood that polarity reversal alone can still damage the cell and the whole stack.

4.9   Cycle Life

Nickel–cadmium batteries are the best of the four main battery system in terms of cycle life and can routinely reach over 1000 cycles. The most important operational factors affecting cycle life are depth of discharge, temperature, and overcharging conditions. Nickel–cadmium batteries can easily reach 1000 cycles for depth of discharge above 80% and several thousand cycles for lower depths of discharged, such as 60%, at which 5000 cycles are possible (Fig. 4.12). Case in point, Ni–Cd batteries have been used in some applications, such as those to power satellites, for over 20 years.

4.10   Self-Discharge

Self-discharge is the one of the most significant disadvantages of nickel–cadmium batteries. At a nominal storage temperature of 20 °C, the rate of self-discharge, or the capacity loss is 10% in the first 24 h and around 20% per month for the first month.

The rate decreases for storage beyond 1 month, but it is still significant for most nickel–cadmium batteries if compared to lead–acid or lithium–ion batteries. If the self-discharge rate is examined not only for the first month, but as an average over several months, it is approximately 10% per month at 20 °C.

At higher temperatures, the self-discharge increases and a battery left in storage, for example in a car, at 40 °C will lose all of its capacity after 2 months (Fig. 4.13). Typically, the rate of self-discharge doubles with every 10 °C (18 °F) of temperature increase.

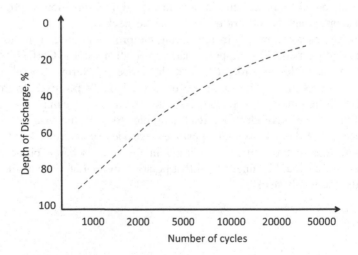

Fig. 4.12 Depth of discharge versus number of cycles showing typical trend (not based on specific battery data, but averaged values for a variety of cells)

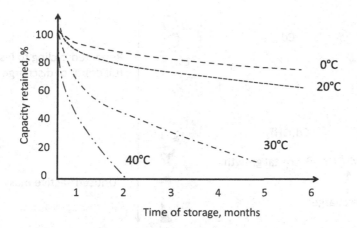

Fig. 4.13 Capacity retained by average Ni–Cd battery versus time of storage for different temperatures. Trends shaped from average data for a variety of cells

The self-discharge does not lead to permanent loss of capacity and recovery of capacity is, in most cases, fairly simple—through a charging process. A useful procedure to maintain full capacity of nickel–cadmium batteries at all times is to use trickle charge simply to offset the self-discharge rate and keep the battery fully charged. If this is not possible, a battery should be stored in cool conditions.

4.11 Memory Effect

Memory effect is an observed loss of capacity when nickel–cadmium batteries are repeatedly discharged to less than 100% depth of discharge and part of active mass remains unused for a period of time. The active mass, Cd on the negative and NiOOH on the positive electrode, which does not participate in cycling reactions on a regular basis undergo growth of crystals due to inactivity. Larger crystals reduce the surface area of contact with the electrolyte and initiate effective loss of capacity. On the negative electrode, in a regular discharge as well as self-discharge reactions, Cd is converted to produce $Cd(OH)_2$ and if there is no timely reverse (i.e., charge) reaction the active mass remains in discharge state, i.e., $Cd(OH)_2$ (Fig. 4.14). In the case of repeated cycling to less than 100 DoD, e.g., 60%, an "unused" portion of the Cd active mass undergoes growth in crystal size (bottom part of the bar in the diagram) due to idleness, while the top or regularly charged and discharged part of the active mass retains small and desirable crystal size. The processes on nickel electrodes similarly lead to memory effect, but to a lesser extent.

The kinetics or rates of reactions leading to Cd crystal growth, while different depending on electrode and design, in general result in permanent formation of large crystals after 6–8 weeks, at which point the reaction is irreversible and active mass of a small crystal size cannot be recovered through a charge reaction. The result is that the reduction in active surface area means effective capacity loss. In the portion of

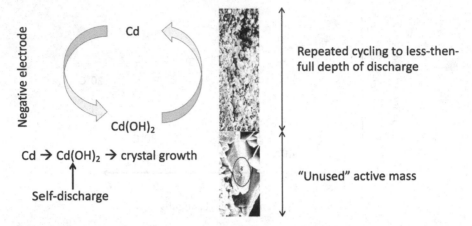

Fig. 4.14 Schematics demonstrating conditions leading to a "memory effect." The bottom bar corresponds to "unused" active mass, in this case metallic Cd

active mass, which is continuously oxidized and reduced the size of the crystals remains small and available for reaction.

Larger crystals have much lower overall surface area exposed to electrolyte and available for reaction; they are effectively not contributing to electrode active mass capacity. It is colloquially said that the electrode "remembers" how much capacity it needs and therefore forgets the rest, resulting in loss of capacity. This has been described in battery jargon as the "memory effect."

Beyond the growth of a crystal, in advanced stages of the process described above, sharp edges of crystals can form and penetrate the separator, causing an electrical short.

Just like the memory effect can affect a battery where a portion of Cd active mass is not regularly discharged and cadmium crystals grow, the memory effect can also occur if a nickel-based battery is left on the charger for days. In this case as well, the cadmium crystals grow and reach sizes beyond which they can be broken down to fine crystals with high surface area. As a result of repeated overcharging a battery appears to be fully charged, but then quickly discharges when connected to a load because of the growth of cadmium crystals and loss of surface area.

A very similar observation can be made for the situation when a battery spends prolonged periods of time in the discharge state without being regularly charged. In this case, the active mass in discharged state, $Cd(OH)_2$ can exhibit crystal growth due to inactivity (Fig. 4.15).

And while memory effect has sometimes been used to assign irremediable disadvantages to nickel–cadmium batteries, there are simple ways to prevent it. A periodic, typically once a month, full cycle of charge and discharge (100% depth of discharge) is sufficient to prevent the memory effect and no further action or service is needed. This is all that needs to be done for batteries used in a standby mode or if a nickel–cadmium battery is oversized for the application and only partial discharge repeatedly takes place during cycling. It should be obvious to the reader that memory

Cd(OH)$_2$ crystals – 1∝m   Cd(OH)$_2$ crystals – 3-5∝m   Cd(OH)$_2$ crystals – 100∝m

Fig. 4.15 Scanning electron micrographs of Cd(OH)$_2$ active mass at different stages of crystal growth due to inactivity over 3 months

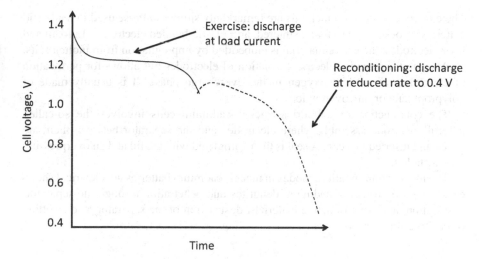

Fig. 4.16 Voltage as a function of time during battery recovery from a condition of large crystal size active mass

effect can also occur if an application is designed with too high cutoff voltage for a battery. This means that 100% depth of discharge can never be reached and memory effect can occur.

Contrary to a popular belief and unfavorable view of nickel–cadmium batteries, the memory effect is not a loss of capacity and not a problem if the nickel–cadmium battery is properly designed for the application. The cell improvements, along with better battery management systems and controls contribute to memory effect being rarely observed. There is in fact no evidence that this was ever a real problem in any application and it has been part of "an urban myth" based on unsubstantiated reports.

Nonetheless, a nickel–cadmium battery should be submitted to a periodic, once-a-month, full discharge, and should not be left on the charger for prolonged periods of time. While the memory effect occurrence has not been a serious problem in practice, engineering solutions have been developed to recover a cell that develops a problem with large crystals. As part of the treatment, a battery is first discharged at a normal regime to 0.9 or 1.0 V. If the battery capacity appears lower due to growth of crystals, a slow, deep discharge is then performed to a cutoff voltage of 0.4 V (Fig. 4.16).

The treatment is expected to break down larger crystals and recover the performance. It is generally assumed that this treatment may not be effective after 3 or more months of crystal growth. In summary, if treated well, nickel–cadmium battery can last for several thousand cycles, a clear advantage over other battery systems.

4.12   Cell Components and Failure Modes

The electrode fabrication methods are remarkably similar to those used in lead–acid batteries: "pocket," fiber, foam, sintered, and plastic-bonded electrodes. In foam and fiber electrodes, the active material is embedded by impregnation from molten salts.

A separator provides electrical isolation of electrodes but allows for permeation of electrolyte as well as oxygen in the overcharge phase. It is usually made of polypropylene or unwoven nylon.

The construction of cylindrical nickel–cadmium cells involves the so-called jellyroll electrode assembly where electrodes and the separator between them are rolled and inserted in a can. A seal is then completed with the lid and an overpressure vent installed.

The most common failure modes in nickel–cadmium batteries are electrical shorts caused by the growth of cadmium dendrites and penetration through the separator, passivation, and wear of active materials, destruction of the separator, and swelling of positive active mass.

Nickel–Metal Hydride Batteries

5

5.1 Introduction

Nickel–metal hydride batteries (Ni–MH) share the same cathode (NIOOH) with Ni–Cd batteries. The anode is a metal hydride material, which enables higher specific energy than Ni–Cd, but results in lower power.

This battery also has good cycle life, but performance starts deteriorating after 200–300 cycles. It is also less prone to "memory effect" than Ni–Cd, but occasional full discharge is necessary every 3 months.

Ni–MH is more sensitive to overcharge than Ni–Cd battery and requires a complex charge algorithm because of more heat generation. Trickle charge settings are preferred because the battery cannot absorb overcharge.

5.2 Principle of Operation

Ni–MH batteries have NOOH cathode and MH anode (M stands for metal). During discharge, MH reacts with OH^- from the electrolyte and is oxidized to metal and electrons, which travel through the external circuit to the cathode and power the load (Fig. 5.1).

Note that metal has an oxidation state of +1 in the hydride and hydrogen is being oxidized on the negative electrode. On the positive electrode, nickel oxyhydroxide is reduced to $Ni(OH)_2$. All reactions are reversed on charge. The overall cell reaction is shown in Eq. (5.1).

$$MH + 2\,NiOOH \quad \rightarrow \quad 2\,Ni(OH)_2 + M \quad E^\circ = 1.31 \ V \qquad (5.1)$$

Similar to Ni–Cd batteries, side reactions evolving H_2 and O_2 become possible in Ni–MH cells near the end of charge and during overcharge (Eqs. 5.2 and 5.3).

© The Editor(s) (if applicable) and The Author(s), under exclusive license to
Springer Nature Switzerland AG 2021
S. Petrovic, *Battery Technology Crash Course*,
https://doi.org/10.1007/978-3 030-57269-3_5

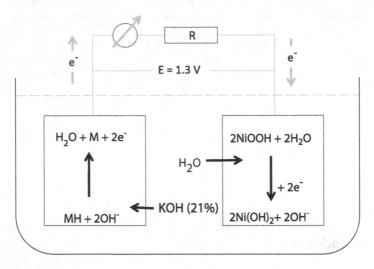

Fig. 5.1 Schematics of a Ni–MH cell and reactions during discharge

Table 5.1 Energy density for metal alloys used for negative electrodes in Ni–MH cells

Alloy	Gravimetric, Ah/kg	Volumetric, Ah/L
AB$_5$	170–290	2200–2400
AB$_2$	360–400	2500–2800

$$\text{Metal hydride electrode} \quad 4H_2O + 4e^- \rightarrow 2H_2 + 4OH^- \qquad (5.2)$$

$$\text{Nickel electrode} \quad 4OH^- \rightarrow O_2 + 2H_2O + 4e^- \qquad (5.3)$$

As in Ni–Cd and LAB cells, oxygen produced at the positive electrode during charge is reduced or recombined on the negative electrode, which is a site for three potential reactions: active mass oxidation, hydrogen evolution, and oxygen reduction.

5.3   Negative Electrode

Two main types of metal hydrides are used in Ni–MH negative electrodes: AB$_5$ and AB$_2$. Candidate metals for these alloys are La, Ce, Pr, Nd, Ni, Co, Mn, and Al for AB$_2$ and V, Ti, Zr, Ni, Cr, Co, Mn, Al, and Sn for AB$_2$.

Despite higher specific energy and energy density (Table 5.1), AB$_2$ alloys are rarely used because of high rates of self-discharge caused by the solubility of vanadium oxide in the electrolyte. The most commonly used AB$_5$ alloys have lower hydrogen storage capacity and lower energy density. They feature low material cost, easy material processing, and less complex electrode preparation.

One more alloy type, A_2B_7, has recently been used for high-energy consumer cells, offering up to 400 mAh/g specific capacity and low self-discharge. The candidate metals for this alloy type are La, Ce, Pr, Nd, Sm, Mg, Ni, Co, Mn, Al, and Zr.

5.4 Charge

The charge process for Ni–MH battery considers the delicacy of the cell chemistry and is in general more complex than for Ni–Cd. Typical charge rates are 0.1–0.2 C.

Control methods for detection of the end of charge include timed charge, voltage drop $(-\Delta V)$, voltage plateau, temperature cutoff, delta temperature cutoff $((-\Delta T)$, and rate of temperature increase. Negative delta voltage (NDV) must be 8–16 mV to match detection sensitivity. However, it is often lower than that for C-rates below 0.5 and at elevated temperatures (Fig. 5.2). Cell aging makes NDV worse and end-of-charge detection more difficult.

Charging Ni–MH cells is even less forgiving at high and low temperatures. Ni–MH cannot be fast charged below 10 °C (45 °F), neither can it be slow charged below 0 °C (32 °F). Consumer chargers are not equipped with temperature sensing, while some industrial chargers are designed to adjust the charge rate to existing temperatures. In general, at higher temperatures, the charge acceptance of Ni–MH batteries is drastically reduced. For example, a battery that provides a capacity of 100% when charged at moderate C-rate at room temperature can only accept 70% if charged at 45 °C (113 °F) and 45% if charged at 60 °C (140 °F). This consideration implies poor battery charging for electrical vehicle packs in warm climates.

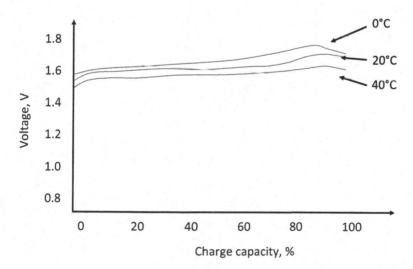

Fig. 5.2 Voltage versus capacity for Ni–MH battery at different temperatures

Lithium Batteries

6

6.1 Introduction

Lithium batteries are the most advanced battery type based on specific energy. Their cell voltage is greater than for other batteries due to highly negative standard reduction potential for lithium of -3.05 V. Additional factor contributing to high-specific energy is low atomic weight of lithium of 7 g/mole.

There are four basic versions of lithium batteries: lithium metal, lithium–ion, lithium polymer, and solid–electrolyte lithium battery. The first two types are based on the nature of anode active material while the other two involve the replacement of liquid electrolyte with the solid. The most well-known and used is lithium–ion battery with liquid electrolyte. Lithium metal battery has theoretically higher energy content, but it is generally unsafe to use, while polymer and other solid–electrolyte batteries have been developed to improve safety for certain applications.

Lithium polymer batteries were developed in response to safety problems related to growth of dendrites in batteries with liquid electrolytes. These cells utilize a thin, polymer membrane as electrolyte and don't contain free electrolyte. The design enables very close spacing between the electrodes and makes these batteries only about 1 mm thick. The batteries are also very lightweight and flexible because the assembly structure is accomplished on thin polymer material, e.g., polyethylene oxide. The polymer electrolyte has lower conductivity than liquid electrolyte and this results in high internal resistance and overall lesser performance. The ionic conductivity of polymer electrolyte increases at elevated temperature and the battery performs better. Some versions of lithium batteries contain a combination of polymer electrolyte and gelled electrolyte for even better conductivity. Both lithium metal and lithium–ion electrodes can be accomplished using polymer electrolyte.

Lithium–ion batteries with different cathode materials have distinctive performance characteristics in terms of capacity, rate or power, cycle life, and other parameters. Some demonstrate superior performance in regard to one of the characteristics, e.g., specific energy, while others show better results for other

S. Petrovic, *Battery Technology Crash Course*,
https://doi.org/10.1007/978-3-030-57269-3_6

Fig. 6.1 Most important
figures of merit for Li batteries

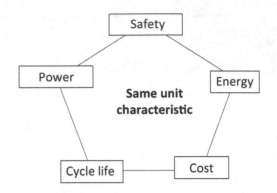

characteristics, e.g., cycle life. Specific versions of lithium battery can effectively satisfy one or two important performance characteristics, but none is superior in all. The five principal figures of merit (Fig. 6.1) are often required for critical applications to have all superior or favorable values, but such a goal has been so far impossible to achieve.

Diverse requirements such as high energy, high power, or high cycle life are not possible with a single lithium battery type and can only be attained with different variants or chemistries. It would be wrong to assume that a single lithium battery can provide, for example, 160 Wh/kg and can also be capable of 30 C current rates or 3000 cycles. These features of superior performance can only be obtained from essentially different batteries that have similar anodes, but with different cathodes and possibly different electrolyte or other battery parameters.

Lithium batteries have good, flat voltage during discharge, low rate of self-discharge, good Coulombic efficiency (i.e., ratio of capacity discharged and charged), and no other serious performance limitations. However, lithium batteries are not perfect. It is important to understand that these batteries are very delicate and require special protection circuits both during charge and discharge. In the absence of protective electronics, which limits voltage and current to safe values, the batteries would very quickly degrade and even lead to catastrophic and hazardous conditions.

The hazardous nature of certain lithium battery conditions is the biggest disadvantage of these batteries. Under some circumstances of unsafe voltage, temperature, or some construction design flaws or defects, lithium batteries can catch fire or explode. For this reason, there are transportation restrictions for the shipment of lithium batteries. They are also expensive compared to other batteries.

Lithium battery is not a fully mature technology. Chemicals used in these batteries are constantly changing, their availability is also questionable and so is the environmental impact.

6.2   Early Lithium Batteries: Li Metal

Lithium metal batteries were first developed in the 1970s and commercialized at the beginning of the 1980s. The first type of lithium battery had lithium metal for anode and molybdenum sulfide as cathode. The cathode material had a layered structure and became a model for subsequent cathodes used in lithium–ion batteries (Fig. 6.2).

The monolayers of MoS_2 are held together by van der Waals forces and such layered structure is significant because it enables lithium insertion and bonding or intercalation between the layers. This type of intercalating compound is prevalent for cathodes in lithium batteries.

The theoretical specific capacity of Li metal can be calculated using Faraday's Law of Electrolysis, as previously demonstrated.

$$\text{Theoretical specific capacity} = \frac{26.8 \text{ Ah}}{6.9 \times 10^{-3} \text{ kg}} = 3884 \frac{\text{Ah}}{\text{kg}} \qquad (6.1)$$

A basic principle of operation of lithium metal battery for discharge reaction involves oxidation on the negative electrode, transport of Li^+ ions through electrolyte and intercalation into cathode material on the positive electrode (Fig. 6.3).

The electrode on the left is lithium metal, shown with dendrite-type outgrowths from the main portion of the electrode and on the right is a representation of the cathode, with layered structure open for Li^+ intercalation. Liquid electrolyte was originally used for lithium metal batteries.

Original lithium–metal batteries had excellent performance regarding specific energy and power but were susceptible to formation of lithium dendrites during the charge process and uncontrolled growth of lithium metal from the negative electrode. The dendrites possess mechanical strength to puncture through the separator and cause electrical short between two electrodes, which results in generation

Fig. 6.2 Model of MoS_2 atomic structure showing monolayers as "honeycomb" sheets of Mo atoms between sheets of S atoms

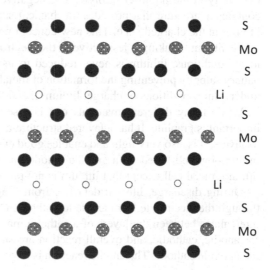

Fig. 6.3 Schematic of
lithium metal battery

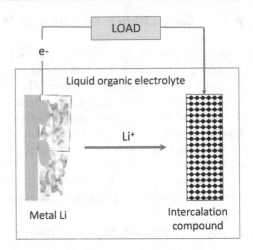

of heat and eventually fire. This type of catastrophic failure was responsible for a
series of incidents in the 1980s with Li–metal batteries catching fire and this
ultimately led to their discontinuation. Eventually, they were replaced with lithium–
ion batteries, which are much safer but also have lesser performance. Recently, there
are indications of the revival of lithium–metal batteries and there are some novel
electrode designs that ensure good specific energy without compromising safety.

6.3   Current Lithium Batteries (Li–Ion)

Lithium–ion batteries are the most commonly used type of lithium batteries. They
were invented in response to safety problems with lithium–metal batteries and its
instability in the liquid electrolyte. The negative electrode in lithium–ion batteries
contains a structure that provides mechanical support and also enables bonding of
lithium in the charged state. This new concept physically mobilizes and chemically
bonds lithium, making it less active in the electrolyte than in a metallic state. Under
ideal conditions, lithium is never reduced to its metallic state and that makes the
battery safer by preventing the formation of metallic dendrites. In practice, however,
under some conditions of charge, lithium metal formation might still be possible.

While many different materials can be used as negative electrodes, the most
important is graphite. It has a layered structure and allows lithium insertion between
its layers. It is also very light and enables good energy density and specific energy. A
lithium–ion cell based on a graphite anode has similar principle of operation as a
lithium–metal cell except that lithium is not present in its metallic form (Fig. 6.4).

During discharge, lithium detaches from graphite, converts to Li^+, and travels
through the electrolyte from anode to cathode. Once it reaches the cathode, Li^+ ion
intercalates between the layers of a cathode material, such as CoO_2. The equations
for anodic, cathodic, and overall reaction are shown below for the case of general
metal oxide cathode. The anodic reaction is fairly simple, whereby elemental lithium

Fig. 6.4 Schematic of Li–ion battery

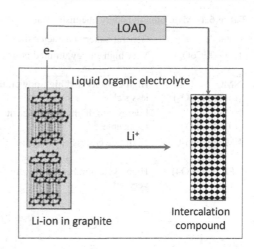

is oxidized to Li^+ and an electron. The cathodic reaction is little bit more difficult to discern, Li^+ reacts with the metal oxide to get partially reduced since fractions are involved.

Negative electrode:

$$Li_xC_6 \;\rightarrow\; xLi^+ + xe^- + C_6 \tag{6.2}$$

Positive electrode:

$$Li_xMO_2 + xLi^+ + xe^- \;\rightarrow\; LiM_{1-x}O_2 \tag{6.3}$$

Cell:

$$Li_xC_6 + Li_{1-x}MO_2 \;\rightarrow\; LiM_{1-x}O_2 \tag{6.4}$$

Reactions in Li–ion battery are reversed during charge, $LiM_{1-x}O_2$ is oxidized on the positive electrode, releasing Li^+-ions, which are conducted through the electrolyte to the negative electrode, to be reduced there and bonded to graphite.

The theoretical specific capacity of a lithium–ion graphite negative electrode is several times lower than for lithium–metal simply because one Li atom bonds with up to six carbon atoms, which present additional weight.

All lithium–ion batteries share a common Li anode, but there are several versions based on the cathode material. Each of these different battery types is characterized by distinct performance characteristics and is therefore used for different applications (Table 6.1).

Lithium cobalt oxide ($LiCoO_2$) is the most commonly used cathode material in lithium–ion batteries. This type of material (or chemistry, colloquially) has a layered structure (Fig. 6.5, left) and is used as power supply for all cell phones and laptop computers. It has high specific energy and moderate cycle life, but limitations such as low discharge currents (<1 C), high cost and concerns regarding safety and cobalt

Table 6.1 Main lithium battery chemistries and intended applications

Cathode	General battery characteristics	Applications
LCO ($LiCoO_2$)	Very high energy, limited power, expensive	Mobile phones, laptops, cameras
NMC ($LiNiMnCoO_2$)	High capacity and high power, thermally less stable	Evs, E-bikes, industrial
NCA ($LiNiCoAlO_2$)	Energy cell, high energy density, thermally less stable	Medical, industrial, EVs (Tesla)
LMO ($LiMn_2O_4$)	High power, less capacity, relatively safe	Power tools, medical devices, powertrains
LFP (LiFePO4)	High cycle number, low energy density, very safe	Portable, stationary

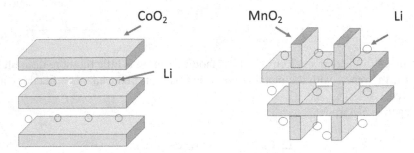

Fig. 6.5 Layered structure of $LiCoO_2$ (left) and $LiMn_2O_4$ (right). Small spheres represent lithium–ions intercalated between the layers of the host structure

availability. The importance of $LICoO_2$ battery is slowly diminishing in favor of mixed oxide materials with better safety and less concerns about raw material availability.

Lithium manganese oxide ($LiMn_2O_4$) has been chronologically the second most popular version of lithium–ion batteries. Its layered, open structure enables high ion migrations rates, which lower internal resistance (Fig. 6.5, right). The battery characteristics include high power density, but at least 20% lower energy density compared with CoO_2. This type of lithium–ion battery has lower cost and performs better at higher temperatures. $LiMn_2O_4$ is safer, more environmentally friendly, and abundant. For this reason, lithium manganese oxide batteries have the capability for high current rates (up to 30 C), both on discharge and on charge. Even at those high currents, this cathode material, also called spinel, has high thermal stability and needs less safety circuitry than the LCO system.

Lithium batteries based on nickel have interesting properties and higher energy density than LCO. By combining nickel, cobalt, and manganese (NMC battery), the expectation is that properties of the resultant cathode and the whole battery can be tuned. A simple approach is that safety can be improved compared to cobalt oxide and power density can be higher because of manganese presence but a tradeoff is that the energy density is lower.

Another three-element cathode material is a combination of nickel, cobalt, and aluminum oxides (NCA) and it demonstrates improved safety, nearly unaffected energy density, but lower cell voltage.

A lithium–ion chemistry that has established an excellent record and interesting properties is lithium iron phosphate technology. These batteries have an excellent safety record, are virtually incombustible, and can better tolerate overcharge and overdischarge conditions as well as high temperatures. Because of this inherent stability of phosphate, these batteries have exceptionally long cycle lives, typically around 3000. This is many times higher than other lithium–ion chemistries and most other secondary battery systems. These batteries have reduced energy density and cell voltage compared to other Li–ion batteries such as LCO.

6.4 Future Lithium System (Li–Air and Li–S)

Two promising cathode materials that could be used in combinations with lithium anode are sulfur and air, both of which are inexpensive and abundant.

Lithium–sulfur battery reaction during charge is depicted in the equation below.

$$2Li^+ + 2e^- + S \;\rightarrow\; Li_2S \quad E^\circ = 2.27 \text{ V} \tag{6.5}$$

This battery has theoretical specific energy significantly higher (>3800 Wh/kg) than all lithium–ion chemistries with metal oxides. Despite the promise, there are still no commercial lithium–sulfur batteries available because of numerous problems (Table 6.2).

Polysulfide dissolution is an especially serious challenge because of the aggregate state change of sulfur, which makes electrode construction difficult, in addition to poor electrode conductivity and the likely loss of capacity (Fig. 6.6).

Table 6.2 Li–S battery problems and their effect on performance

Problem	Effect on performance
Li dendrite growth	Electrical short
Low electronic conductivity	Voltage drop, low capacity
Low mass loading of S	Low capacity
Polysulfide dissolution	Active mass loss, capacity fading

Fig. 6.6 Diagram of polysulfide dissolution during charge/discharge processes

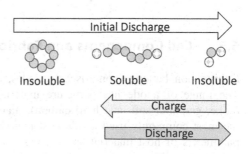

Fig. 6.7 Schematics of a Li–Air battery

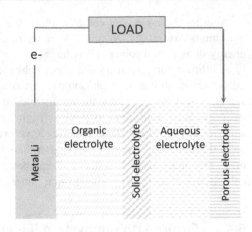

Lithium–air batteries have theoretical specific energy of more than 5000 Wh/kg—the highest of all battery systems. The main characteristic of this battery is that cathode active material is not stored in the battery; therefore, the cathode is lighter and smaller because volume and weight are eliminated. The reactant on the cathode is oxygen from air and it is supplied through openings in the cathode when under load (Fig. 6.7). It is challenging to operate a system open to air, maintain electrolyte in the pores of electrodes, and develop efficient oxygen electrocatalyst. To prevent the reaction of organic electrolyte with oxygen, the cell is accomplished by having a solid electrolyte between organic electrolyte on lithium side and aqueous electrolyte on oxygen side.

During the discharge, lithium is oxidized with oxygen and during the charge oxide decomposes releasing oxygen.

$$2Li^+ + 2e^- + O_2 \ \rightarrow \ Li_2O_2 \quad E^\circ = 3.14 \ V \tag{6.6}$$

These reactions are relatively simple; however, the biggest challenge is to prevent unwanted reactions between lithium and oxygen. Most of the results from lithium–air batteries research show very quick degradation in capacity after only a few cycles.

Despite slow progress, both Li–S and Li–Air systems have been extensively investigated, holding hope that a better battery system will be realized.

6.5   Cell Components and Fabrication

Lithium–ion batteries comprise electrodes, electrolyte, separator, and packaging. The range of anode host structure materials includes various forms of carbon (coke, graphite, soft, and hard carbon), titanium dioxide, and silicon. Graphite is the most commonly used anode host material in all practical batteries. Important parameters of host material are spacing between the graphite layers and volume

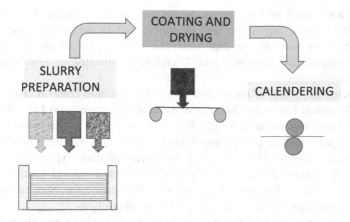

Fig. 6.8 Diagram of electrode preparation from active mass slurry through coating on current collector

expansion with lithium intercalation. For graphite, the spacing is 0.335 nm and volume expansion is about 10% for fully charged anode.

Soft carbon is less ordered and has more variable spacing between the planes. Hard carbon is noncrystalline and even less ordered, with spacing between the layers of more than 0.38 nm. Hard carbon is also most resistant to volume changes upon intercalation of lithium and therefore has better prognosis for good cycle life. Two novel anode materials are titanium dioxide and silicon. Both have been extensively studied to overcome problems with volume changes upon intercalation of lithium.

The electrodes are prepared through a process of coating thin metal foil with a slurry containing active material, binders, and conductivity enhancers. The anode active mass is one of the materials described above, typically graphite; and cathode active mass is one of the materials discussed in the previous section, such as lithium cobalt oxide. Active material, binder, and conductive agent are mixed together in specific mass ratios, followed by a tape-casting process on the current collector (Fig. 6.8). Finally, during the so-called calendaring step, the porous electrodes are compressed by driving them between two massive cylinders.

Electrodes are subsequently cut into desired size and connection tabs attached, before forming assemblies by "winding" procedure with the separator, whereby one or more pairs of electrodes are stacked with a separator in between.

Next, for cylindrical cells, the assembly is rolled and inserted into a stainless steel or aluminum can. For larger cells and for most portable applications as well, a "pouch cell" is formed by inserting rectangular, flat assemblies into a pouch, which is polymer-coated aluminum. After vacuuming and injection of electrolyte, the pouch is capped and sealed, then the battery is ready for formation charge and use. In the formation charge, lithium from the cathode active material is oxidized and moved through electrolyte (as Li^+) for the first time into the anode structure, where it intercalates.

A separator is used between the anode and the cathode to enable close proximity of electrodes, which minimizes the resistance, and to prevent electrical shorts between the electrodes. At the same time, the separator has to enable electrolyte penetration that ensures ionic transport. Therefore, a separator has porous structure, with submicron pores.

The electrolyte is one of the several proprietary organic compounds in organic solvent. The electrolyte is nonflammable and it should have good Li^+ ion conductivity and large electrochemical stability window (>5 V). The most common electrolyte used is lithium phosphorus fluoride, $LiPF_6$. It is dissolved in a combination of solvents, typically ethylene carbonate mixed with dimethyl carbonate, diethyl carbonate, or methyl carbonate. Sometimes electrolyte is in gel form to prevent formation of lithium dendrites. Recently, developments have been underway to synthesize ionic liquids as replacement for liquid electrolyte.

In summary, lithium cell fabrication includes the following steps: slurry coating, calendaring, winding, assembly, cap and seal, and formation.

6.6   Charging

Lithium battery charging proceeds through a two-stage process (Fig. 6.9).

The first stage of this process is the constant-current stage, which takes a little over 1 h and during which voltage gradually increases from open circuit of around 3.5 V to a preset value between 4.1 and 4.2 V for most lithium batteries. It is critical for proper battery performance that this value is never exceeded, to prevent battery damage and hazardous conditions. At the same time, charging to less than maximum voltage diminishes capacity, for example, charging to 4.1 V instead of 4.2 V reduces

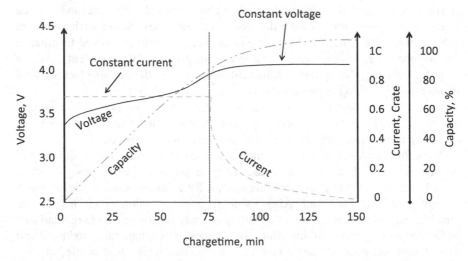

Fig. 6.9 Voltage, current, and capacity versus approximate time in minutes for lithium–ion charging process

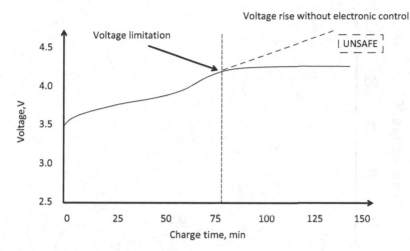

Fig. 6.10 Battery voltage versus charge time depicting surpassed limitation voltage and unsafe conditions

capacity by 10%. For this reason, a charger must have the capability to discontinue constant current charging when the exact voltage limitation value is reached. During this first stage of charging, the state of charge or capacity increases gradually and reaches 70–80%.

After the limitation voltage is reached, the charger (i.e., power supply) switches from constant current to constant voltage mode, at the limitation voltage. While voltage is now kept constant, the current exponentially decreases to a value of 3% of the starting current—at which point the charging is complete. The second charging stage, constant voltage, takes at least as long and normally twice as long as the constant current stage. The complete charging process is typically completed in 3 h after full capacity is reached at the end of the second stage.

Ratio of times between two stages is not easy to manipulate, i.e., increasing the charge current in the first stage cannot shorten the overall charge time by a significant amount. The maximum voltage may be reached faster with higher current, but the second stage will then take longer. In the end, around 3 h is the time necessary for the full charge of lithium batteries and it is impossible to fully charge a battery in the first stage alone. This is an especially important consideration for fast electrical vehicle charging, where high rates of 5–6 C are desirable.

Lithium batteries should not be allowed to reach the overcharge state (Fig. 6.10). If maximum (or limitation) voltage is exceeded, rate of Li^+-ion transport to the negative electrode surpasses the rate of intercalation. Under those conditions, instead of intercalation reaction a deposition (or plating) of Li metal can occur and create dangerous growth of Li metal crystals from the negative toward the positive electrode. The commonly used term for metalic Li branch-like structures, is dendrites. If not suppressed, dendrites propagate toward the cathode by puncturing a separator, causing an electrical short; and occasionally dangerous heat evolution and fire.

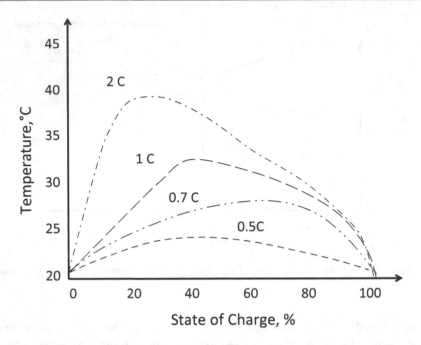

Fig. 6.11 Cell temperature as a function of state-of-charge at different C rates. Particular curves are adapted from a variety of sources and are not representative of any particular battery or experiment

To prevent dangerous conditions resulting from Li dendrite growth, Li–ion batteries are equipped with an internal protection circuit, which disconnects the cell from the charger. Other types of protection, such as monitoring cell temperature, are also common. Cell temperature monitoring typically activates a separator shutoff to stop battery reactions if a dangerous temperature is reached. Heat generation and temperature increase in Li–ion cells depending upon current, rate of change of voltage over time (dV/dt), and on the difference between cell voltage and open-circuit voltage. This complex interaction can lead to an initial increase in temperature, followed by gradual heat dissipation and temperature decrease (Fig. 6.11).

Risk of Li plating (instead of intercalation) is the reason that continuous toping or trickle charging is not recommended for lithium batteries and it should only be used intermittently to compensate for the self-discharge.

6.7   Discharge

The open-circuit voltage in Li–ion batteries shows fairly linear dependence on the state of charge. However, it is not recommended to assume that open-circuit voltage is an accurate indication of the state of charge. The relationship between the two battery parameters is more complex and open-circuit voltage only gives a rough indication of the battery state of charge.

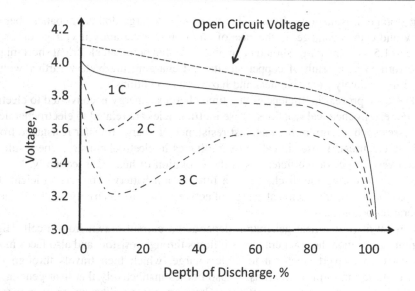

Fig. 6.12 Conceptual trends for voltage versus depth of discharge for Li–ion battery and different C rates. The trends in this graph reflect only one type of Li–ion battery

The shape of the voltage curve for 1 C current shows typical behavior, with initial drop and then gradual decrease until the end of discharge is nearly reached and the voltage sharply drops (Fig. 6.12).

The shape of voltage curves for 2 C and 3 C currents is more unexpected. There is a depression in voltage in the first third of discharge, then voltage increases, stabilizes, and decreases again near the end of discharge. It is important to understand that this particular performance behavior or shape of voltage curves is not universal for all lithium–ion batteries. This is just one example for a battery used in automotive applications. Other batteries of different size and construction may have different heat generation and dissipation properties and their voltage profiles would be different.

The limitation or "cutoff" voltage at the end of discharge is also a very sensitive issue for lithium–ion batteries. In order to prevent overdischarge, most cells are disconnected, that means that discharge is completed when voltage reaches 2.7–3.0 V (typically 3.0 V). The safety protection circuit is ordinarily part of a battery and it disconnects the current if the battery is inadvertently discharged below 2.50 V per cell. If this voltage is reached, the cell is considered unserviceable, it is put to sleep and regular recharge is not possible. Instead, a cell might be recoverable, but only through a special "wake-up" protocol.

A complex situation may develop during prolonged storage of lithium–ion batteries. Most manufacturers ship batteries with about 40% state of charge to compensate for some self-discharge during storage. The safety circuitry is not activated during storage and cannot protect against deep discharge below 2.5 V until the first, even brief, charge cycle. Before this activation step, in the case of very long storage, it is theoretically possible that self-discharge drains the battery below 2.5 V and causes damage. In the case of any doubt or certainly if voltage drops to close to 1.5 V—a recharge should be avoided. At that point, an electrical short might have formed as a result of copper shunts and charging the battery further would cause immediately excessive heat and hazardous conditions.

There are process inefficiencies any time chemical energy is converted to electrical. For electrochemical reactions, these inefficiencies are related to electrochemical processes such as current flow against resistance or charge transfer resistance from ionic to electrical. These are called overvoltages in electrochemistry. The result of these overvoltages or inefficiencies is the evolution of heat. Consequently, heat is also evolved during the discharge of a lithium–ion battery. The less efficient the process for converting chemical energy of active masses to electricity, the larger heat generation becomes.

It is intuitive that heat generation depends on current drawn from a cell. High current causes more heat generation as it flows through resistors and also faces more resistance to convert electrical to ionic charge, which then travels through the electrolyte in the form of Li^+ ions. It also comes instinctively that heat generation should be proportional to the difference between the equilibrium or open-circuit voltage of a battery and the operational voltage in the moment of observation. The higher this voltage difference is, the greater the resistance the current faces as it flows through the system, both in the form of electrons through the external circuit and in the form of Li^+ ions through the electrolyte. Heat generation in a lithium battery is given by the equation below, where E is the voltage at the time of observation, I is the current, and T is the temperature.

$$Q = I\left(E_{oc} - E - T\frac{\partial E_{oc}}{\partial T}\right) \tag{6.7}$$

Especially resistive is the last step in the intercalation mechanism where Li^+-ions drift toward their intercalation sites inside the crystalline structure of the host cathode compound, for example, CoO_2. At that point, near the end of discharge, most of the available sites for lithium intercalation have already been occupied and the incoming Li^+-ions have to meander in somewhat torturous path to find available sites for reduction by the electrons from the outside circuit. This causes resistance and heat generation.

It is also evident from the equation that heat generation is also dependent on the rate of voltage change with time. From examination of familiar voltage trends and comparison with the temperature curve (Fig. 6.13), it becomes apparent that in the beginning of discharge the voltage drops steeply and in the same period temperature sharply increases. Then, for the majority of discharge, voltage is stable, or it slowly declines, while the temperature gradually rises. Finally, in the last stage of discharge, voltage sharply drops, while temperature abruptly rises. It can be said that the temperature curve is roughly an inverted image of the voltage curve and our reasoning supports that.

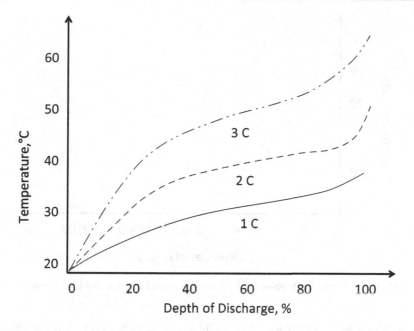

Fig. 6.13 Conceptual trend of temperature changes in Li–ion cell during discharge, for different C rates

It is also clear that temperature depends on discharge current and hazardous conditions can ensue along with high currents if battery internal resistance causes excessive heat generation and temperature increases beyond safe limits. For 3 C currents, for example, the temperature can rise over 50 °C. It should be noted, however, that different battery types and different chemistries exhibit different behavior and individual temperature profile must be understood for each specific battery. Most cells have protection circuits that will disconnect and disable the battery if unsafe temperature is reached.

6.8   Cycle Life

Lithium–ion batteries follow a similar pattern of cycle life dependence on the depth of discharge as other batteries (Fig. 6.14).

A familiar observation is that depth-of-discharge significantly influences the number of cycles and that higher depth of discharge on each cycle results in lower cycle life. For 100% DoD, most lithium–ion batteries have cycle life of 300–500. Lower depth of discharge means extension of cycle life and very shallow cycling can dramatically extend cycle life to several thousand. While this sounds like a simple way to better utilize a battery, it should be pointed out that a design that includes lower depth of discharge also means that a battery would be oversized compared to

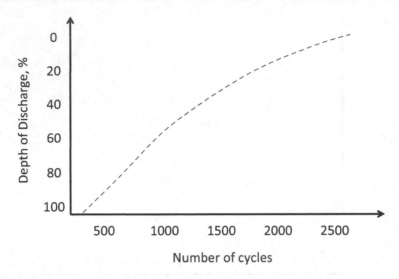

Fig. 6.14 General trend for depth-of-discharge versus number of cycles for Li–ion batteries

demands of a load, which would certainly result in higher cost. A design would consider if higher cost may bring a benefit of less frequent battery replacement.

Contrary to nickel-based batteries, lithium batteries do not need periodic full discharge. In general, it is better for the battery life if they are not discharged too deeply and are fully charged more often.

Capacity of a battery, compared to first cycle capacity, is a measure of the state of health of the battery and usually a good indication of the number of cycles up to that point. The capacity steadily decreases, typically starting after around 200 cycles. Eighty percent of the initial capacity is usually considered end of life for many critical applications, such as in portable electronics or automotive batteries. At that point, most lithium batteries would have reached close to 500 cycles.

The most popular, lithium cobalt oxide and lithium manganese oxide have cycle life typically well below 1000. For 100% DoD, these batteries reach normally 300–500 cycles (Fig. 6.15, dashed line).

There is one lithium–ion chemistry, the iron phosphate, that exhibits exceptionally high cycle life. It is shown in the graph for one particular manufacturer, that cycle life is between 2000 and 5000, depending on DoD. This is an extraordinary improvement compared to most common lithium–ion chemistries.

It should also come as no surprise that cycle life depends on the nature of cathode material. With a different cathode material, such as iron phosphate, the cycle life significantly increases because the material is more stable than, for example, cobalt oxide and remains stable over a large number of cycles. This unique stability also contributes to much improved safety. Since the material is more stable, there is less risk of violent reactions involving lithium or other components. However, this lithium–ion chemistry has lower specific energy.

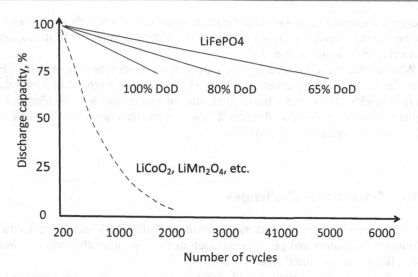

Fig. 6.15 General trends for Li–ion battery capacity change versus number of cycles

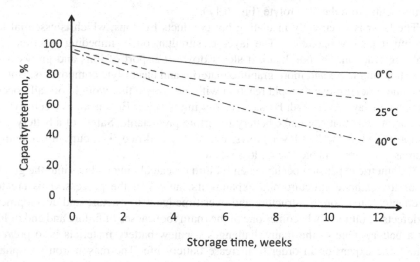

Fig. 6.16 General trends of capacity retention versus time for Li–ion batteries stored at different temperatures

6.9 Self-Discharge

Compared to nickel batteries, lithium batteries have very favorable self-discharge rates. Except the first day self-discharge, the rates are typically 1–2% per month and are temperature dependent (Fig. 6.16). The higher the temperature, the higher rate of self-discharge reaction. The trend is similar to all other batteries. The rate of self-discharge is much faster at +40 °C and capacity retained is about 50% after 3 months. The best storage conditions are at low temperatures, for example, at 0 °C the capacity retained after 3 months is over 95%.

When evaluating self-discharge, it is important to consider the state of charge. For example, at nominal temperature and 40% state of charge a typical battery would retain over 95% of its capacity but at 100% state of charge only 80%. At temperature higher than nominal, for example at 40 °C, the capacity retention values for 40% and 100% SoC are typically 85% and 65%.

6.10   Operational Challenges

After examining performance factors for lithium batteries, it is important to delineate operational challenges and phenomena, such as the role of solid–electrolyte interphase, lithium safety, and thermal runaway.

A solid–electrolyte interphase (SEI) layer on carbon surface is created in the first (or formation) cycle through the interaction between carbon particles and components from the electrolyte (Fig. 6.17).

The layer is electrically insulating but conducts Li^+-ions, which is essential for continued battery operation. The layer is simultaneously impeding reaction and enabling reaction. On one hand, it slows down transport of ions and grows over time, trapping more and more graphite material and electrolyte components. But, at the same time, it prevents rapid reaction with electrolyte that would normally occur when a new layer is created. By slowly growing in size, SEI consumes electrolyte at a moderate rate that can make battery life more predictable. Safety of a battery gets compromised as the SEI layer grows, cyclability and rate (i.e., current capability) decrease, and irreversible charge loss occurs.

Volumetric expansion occurs when lithium repeatedly intercalates into the graphite of the cathode structure and expands its lattices in the process. This creates mechanical stress on the structure and over time leads to fracture and development of defects. It ultimately becomes one of the main root causes of failure and end of life for a battery. One of the main challenges for new battery materials is to prevent volumetric expansion in order to increase battery life. The reason iron phosphate chemistry exhibits high number of cycles is that this material is stable and doesn't undergo damaging expansion and contraction in every cycle.

Fig. 6.17 Schematic of the formation of solid electrolyte interphase layer on Li graphite electrodes

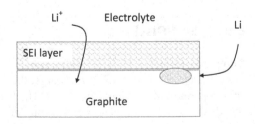

As previously discussed, if a battery is properly designed and manufactured without defects, the thermal balance is stable, and the battery never reaches unsafe temperature levels. However, there are three external factors that contribute to heat generation and dissipation imbalance, which can lead to unsafe rise of temperature and thermal runaway. These factors are mechanical failures, electrical abuse, and external temperature. The onset of a thermal runaway can be triggered, for example, by electrical abuse in the form of uncontrolled overcharge, causing excessive heat evolution. Once temperature in a cell increases beyond a safe value, the rates of all reactions increase according to the Arrhenius equation.

$$k = Ae^{\frac{-E_{act}}{RT}} \tag{6.8}$$

where k is the rate constant, T is temperature in K, A is a constant for every chemical reaction, E_{act} is the activation energy for the reaction, and R is the universal gas constant.

All reactions, desired and undesired side reaction became faster and more heat is generated, which has to be dissipated out of the cell. If cell design and components used are not capable of dispersing a higher quantity of heat, then cell temperature will further increase, triggering additional reactions. Three characteristic temperatures are onset, acceleration, and trigger temperature. Once the trigger temperature has been reached, an uncontrolled reaction starts and thermal runway is inevitable. It results in pressure increase as well, mechanical damage to a cell, ingress of oxygen or moisture, and finally fire and explosion (Fig. 6.18).

A more detailed look at the mechanism reveals that first stage of the overall runaway temperature process is disintegration of the solid–electrolyte interphase layer. This layer, while somewhat impeding battery reactions, also serves as a protective layer, preventing violent reaction. But as a result of increased temperature (~80 °C), the SEI layer breaks down. The electrolyte now reacts unhindered, and at a higher temperature, with carbon from graphite, resulting in uncontrolled exothermic reaction that further increases the temperature.

As the temperature increases to close to 100 or 110 °C, the heat is now excessive, and it causes the breakdown of organic solvent, which results in the release of flammable gases, such as ethane and methane. At this temperature, the flashpoint of hydrocarbon gases is exceeded, but there is no fire because there is no free oxygen inside the cell.

However, gas evolution also results in pressure buildup. Modern lithium–ion cells have safety vents, which open when a certain pressure is detected and release the gases. It is possible that upon the release into the atmosphere gases may briefly burn in contact with oxygen from air. The function of a safety pressure valve is critical, otherwise an explosion and cell rupture could occur.

As temperature further increases, at 135 °C there is a built-in safety feature when the separator melts, preventing ionic contact and stopping the reaction. However, at this point, the heat evolution is uncontrolled and, as it rises to around 200 °C, causes disintegration of metal oxide, for example, cobalt oxide from the cathode. This is also an exothermic reaction and heat is generated. But this has even more serious

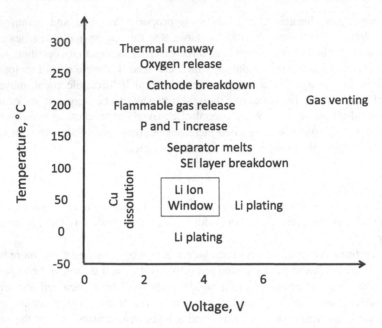

Fig. 6.18 Conceptual reactions, temperature, and voltage for thermal runaway event in Li–ion batteries

implications as the breakdown of oxide releases oxygen and the hydrocarbon gases from electrolyte start burning, causing further temperature increase and fire, leading to a full, catastrophic failure of the battery.

It is obvious from this description that one critical step in the thermal runaway sequence is the decomposition of metal oxide. For cobalt oxide, decomposition occurs by the time temperature inside the cell reaches 200 °C. This relative instability of the most popular cathode material, CoO_2 is the reason for safety concerns with lithium–ion batteries. Some other cathode materials, e.g., iron phosphate decompose at a higher temperature, around 300 °C. This makes an iron phosphate cathode more stable, resulting in better cycle life and more importantly it makes it much safer. The likelihood of internal cell temperature reaching 300 °C and causing decomposition of iron phosphate is extremely small. As that critical step in the thermal runaway is eliminated, the safety concerns have lessened.

Thermal runaway for lithium–ion batteries is a topic of extensive testing. Accurate temperature monitoring, accelerated rate calorimetry, and tomography are some of the techniques used.

Epilogue

The field of batteries, the scope of that field, the variety of battery systems, the diversity of application areas, and the options for advancement are much broader topics than what has been presented in this crash course. While the course provided an initial understanding of batteries, it was meant to unlock opportunities for further learning and encourage exploration of the future of batteries. Beyond the four historically important battery systems used in the crash course to elucidate the most fundamental principles of batteries, the foundation of knowledge will enable a rapid understanding of additional systems such as metal-air, flow batteries, batteries based on a sodium anode, and the most promising advanced battery combinations, such as lithium-air, lithium-sulfur, and all solid-state batteries. If this crash course managed to remove the vail from the enigmatic world inside a battery, then the next step should be to venture beyond and consider what the future might bring and how the boundaries of improvement can be stretched. Ultimately, progress in the field of batteries will continue to happen and the only question is who is going to be part of it.

Index

Printed in the United States
by Baker & Taylor Publisher Services